Living Off the Land

MODULE F

MODULE F

Living Off the Land

Land ho! Land covers less than one third of the earth's surface. Good farm land covers even less—and it comes with lots of bugs. In this module, you'll find out how farmers and scientists are working together to grow more food from the land, while fighting off an ever-growing army of insects.

CHAPTER

1 Planet of Plants

Can you smell danger in the air? Some plants may use chemical odors to warn their comrades of approaching danger.

CHAPTER

2 Growing Plants

How is a flower like a factory? Both manufacture products. For flowers, the products are life-giving seeds.

CHAPTER
3 Gardening Plants

Green thumbs are as close as your fingers. Know-how and hard work can produce gardens that are bursting with life and food.

In this module

At the end of the book

CHAPTER

1 Planet of Plants

I can't smell anything yet!

Discover Activity

What do odors tell you?

Your classmates should stand in different parts of the classroom. You or your teacher will open a bottle containing a liquid. As soon as your classmates smell the substance, they should raise their hands. Wait until all hands are raised. Repeat with a new substance.

For Discussion

1. *How did the odors travel throughout the room?*

2. *What was your reaction to each odor?*

LESSON

1.1 Stalk Talk

Do plants communicate with each other?

Something's happened. You've shrunk to the size of a peanut. You have wings and long, floppy antennae. You are teetering on a potato leaf in a vegetable jungle. Corn, squash, and potato plants tower over you. As you can see, it's lunchtime. And the potato plant just happens to be your favorite snack. You're a Colorado potato beetle! Farmers and gardeners don't like you very much.

The garden is hopping—like a school cafeteria. You hear the crunch of insects chewing, and the hum of "conversation"—buzzing, chirping, and chattering. A mingling of smells is in the air—the scents of tasty foods, a whiff of sweet perfume.

You move on to the next potato plant. But its leaves taste *bad*. "Bleech!" you say and spit it out. Sounds weird, but sometimes you feel like the plants get the word that you're hanging around. They seem to make themselves taste awful before you have a chance to chow down. You squat on a leaf and scratch your head.

What kinds of communication do you see in this cartoon?

Plant Talk Hypothesis

Scientists are scratching their heads, too. They're asking themselves: Do plants really "talk" back and forth? Do they communicate and warn each other, as this cartoon suggests? Does one plant "say" to its neighbor, "Hey, the bug that likes potato leaves is heading your way"?

Scientists have found that certain plants have a way of protecting themselves from attacking insects. These plants have "buzzers" that go off when insects take nips out of them. When the chomping begins, an alarm signal travels through the plant. This signal causes the plant to make chemicals that are bad news for the munching insects. The anti-insect chemicals poison the insects or stunt their growth.

Now scientists are testing a related idea about alarm signals—the "plant talk" hypothesis. Some scientists think that a plant can do more than just protect itself. They think a plant can also send out an alarm signal that warns its neighbors of danger.

Describe what you think is happening in this cartoon.

The hypothesis goes like this: After an insect chews a leaf, the injured plant releases a chemical that floats through the air. Remember that in the Discover Activity you learned how chemicals travel through air. You also found out that different smells cause you to respond in different ways. For example, some of the odors might have caused you to turn away or hold your nose.

According to the hypothesis, the chemical signal given off by the injured plant may cause a response in other plants. The chemical may act as a warning that shouts "Pests at lunch!" With early warning, nearby plants can start to make anti-insect chemicals *before* they get nipped by the pests.

A chemical in the sagebrush plant on the right spreads through the air in this jar. The chemical may act as a signal to the tomato plant nearby.

In the laboratory, scientists found evidence to support the plant talk hypothesis. Sagebrush plants contain a chemical that some scientists think is a powerful alarm signal. The scientists tested their idea by doing the experiment pictured here. They put sagebrush plants and tomato plants together in air-tight jars. Guess what happened? Tests showed that the tomato plants began to make anti-insect chemicals. The scientists concluded that the alarm signals given off by sagebrush traveled through the air and caused this response in the tomato plants. To further test their hypothesis, the scientists put tomato plants in the air-tight jars *without* sagebrush plants. These tomato plants did not produce the anti-insect chemicals.

Do scientists need more evidence for the plant talk hypothesis? Of course they do! Remember, they did their tests in the laboratory. Plants don't normally grow in jars. Now scientists are doing new experiments that are testing the hypothesis out-of-doors in the plants' own territory.

Plant Signals

Scientists don't know for certain whether chemical signals travel from plant to plant. But they are quite sure about one thing. Chemical signals do travel *inside* a plant—from one part of a plant to another part.

When you plant an apple seed, what causes the seed to crack open? What causes a tiny root to come out of the seedcase first, and push downward into the soil? What causes a young shoot to poke through the ground and grow upward?

Chemicals called hormones control all of these activities. A **hormone** is a chemical messenger with a precise list of instructions.

A single plant contains a crew of hormones, each with its own assignment. One hormone is in charge of making roots grow downward. Another directs a stem to grow toward sunlight. When the time is right, other hormones make an apple ripen and turn red, or a maple leaf change color.

Hormones control all the activities connected with the growth and development of plants, such as this apple tree. In turn, the long days of summer, the cold nights of winter, and other signs from nature control hormones. Environmental signals—which you call the changing seasons—help control how hormones are made and released within plants.

Hormones control the growth of an apple seedling.

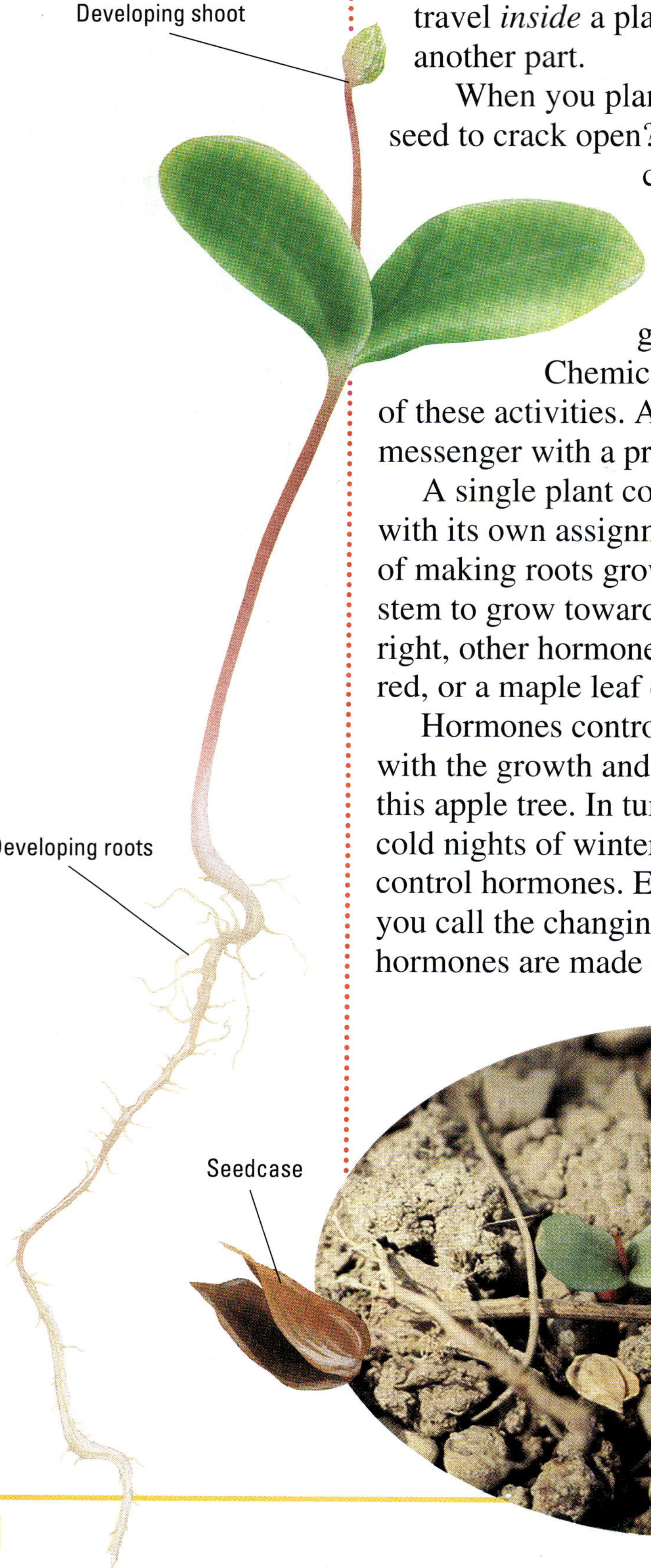

Most hormones are produced in one part of a plant, but they have their effect in another part of the plant. So, how do they get the job done? Hormones move through tubes that carry food throughout a plant. And hormones are constantly on the march inside a plant. For example, one hormone is in charge of making a plant bend toward light. This hormone is produced in the very tip of the stem, but it travels down into the stem with its instructions.

As chemical messengers, hormones don't waste words. A tiny amount of a hormone can bring about a big change in a plant. For example, an extremely small quantity of a kind of growth hormone causes the shoot and root of an apple seedling to grow longer.

Besides controlling activities in different parts of the plant, hormones sometimes control each other. When it's time for a plant to reproduce, the hormone in charge of fruit-ripening has control over the hormone that keeps the fruit green.

Hormones control the ripening of the apples on this tree.

Checkpoint

1. Why must scientists design experiments that will test the plant talk hypothesis out-of-doors?
2. List four activities of plants that are controlled by hormones.
3. **Take Action!** Put a very ripe banana and a green tomato in a covered shoe box. Wait a few days and open the box. Observe how a hormone given off by the banana has ripened the tomato.

Activity

Putting Down Roots

People can use plant hormones to help plants grow. Try this activity to see how rooting hormones work.

Picture A

Picture B

Picture C

Gather These Materials

cover goggles
2 clear plastic cups
vermiculite
pencil
mature plant (coleus)
rooting hormone
2 clear plastic bags
masking tape
water

Follow This Procedure

1. Make a chart like the one on the next page. Record your observations in the chart.
2. Put on your cover goggles.
3. Fill your 2 cups with moist vermiculite.
4. Use a pencil to poke 2 holes in the vermiculite in each cup. The holes should be up against the inside wall of each cup. (Picture A)
5. Take 2 small cuttings from the coleus plant. Insert a cutting in each of the holes in one cup. You should be able to view the tips of the cuttings through the side of the cup. (Picture B)
6. Pack some more vermiculite around the cuttings to hold them in place.
7. Take 2 more cuttings from the coleus plant. Have your teacher dip the ends of the cuttings in the rooting hormone.

8 Plant these cuttings in the second cup. Follow the same procedure you used for the first cuttings.

9 Place a clear plastic bag over each cup. Loosely tape the bag in place. (Picture C)

***Predict:* Which cuttings will have greater root growth?**

10 Keep your cuttings in a warm place with plenty of light. Remove the plastic bags to water the plant cuttings as needed.

11 Check your cuttings every day. Look for root growth. Record your observations.

Record Your Results

Observations		
Date	Cuttings with root hormone	Cuttings without root hormone

State Your Conclusions

1. How did the root growth differ in your 2 cups?
2. Do you think that the root hormone helped the plant grow roots? Explain.
3. How would using rooting hormones be helpful to farmers or gardeners?

Let's Experiment

Do you think that some types of plants grow roots faster than others? Use what you know about scientific methods to find out.

LESSON

1.2 The Talking Planet

What's the flower telling the butterfly?

While scientists continue to explore the idea of plants "talking" to other plants, they know that plants do "talk" to insects. Plants produce chemical signals that contain important messages.

A smell is a message. Flowers attract insects with fragrances that say, "Here's food for you." Butterflies are attracted to flowers that smell sweet. Beetles are attracted to flowers with spicy scents.

A bad taste is another message. Some flowers contain a poison called pyrethrum (pī rē′thrəm). Its message is blunt: "Bug off!" An insect gets the message after its first taste.

Take a look under the ground at the roots of this alfalfa plant. Find the nodules where bacteria live.

Signals in the Soil

Insects aren't the only organisms that get messages from plants. Tiny bacteria in the soil receive messages from plants called legumes (leg′yüms). And the bacteria actually "talk" back!

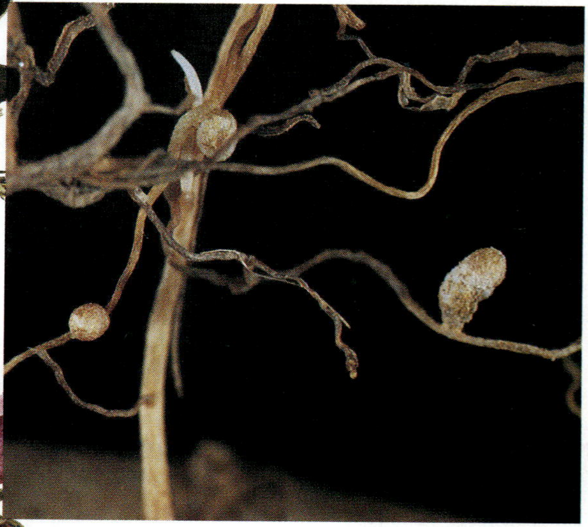

Legumes are plants such as soybeans, peas, and alfalfa. These plants can change nitrogen gas in the air to forms of nitrogen that plants need for growth. Well, the legumes don't exactly do this on their own. The plants provide homes and food for bacteria called *Rhizobium* (rī zō′bi um). But the bacteria do the work.

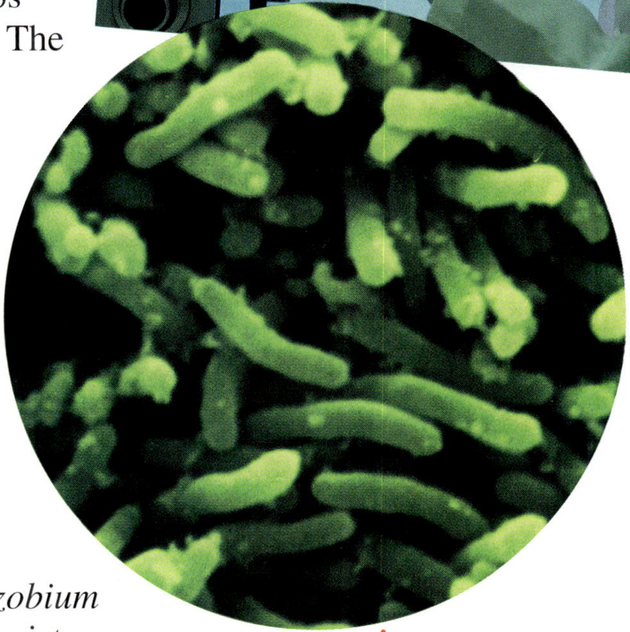

As the pictures show, the legumes construct little bumps called nodules on their roots. The nodules become a home and factory for the bacteria. The bacteria live inside the nodules and feed on sugars made by the plant. In return, the bacteria make the nitrogen compounds that the plants need for growth. So, as you can see, both the plants and the bacteria benefit from this partnership.

How do legumes and *Rhizobium* get together? Molecular biologist Sharon Long discovered that alfalfa plants and *Rhizobium* "talk" to one another. Alfalfa starts the "conversation" by sending out a chemical called a flavonoid. The chemical puts out a call to all *Rhizobium* in the soil: "Come to work in exchange for room and board."

Scientists are using microscopes and other tools to study these Rhizobium *bacteria, which are found inside alfalfa nodules.*

When the *Rhizobium* receive the message, they send their own chemical reply. They release a chemical into the soil that carries this message to the alfalfa plant: "We'll be there soon. Start making our nodule homes." Long and her co-workers think the conversation continues. They are doing experiments to find out what the alfalfa says next.

Each antenna of the male silkworm moth may have 70 branches. Each branch has thousands of hairs that detect tiny amounts of female sex pheromones.

Talking Long Distance

As you learned, a hormone works *within* a living thing, where it carries messages to certain parts of the organism. A **pheromone** is a chemical that communicates information between two or more members of the same species. You might say that pheromones are long-distance hormones.

Pheromones are especially important in insect communication. Some insects, like butterflies, live by themselves. Others—like ants, aphids, and honeybees—live in large, crowded communities.

Without pheromones two tiny insects might *never* find each other in the thick grass. In many species, a female releases a sex pheromone that causes a male of her species to stop in his tracks. He picks up the signal with long, feathery antennae like those in the picture. The antennae can detect very tiny amounts of the female pheromone. When a male detects the scent, he will follow the fragrant pheromone path directly to the female. The male might then release his own pheromone to let the female know that she is about to be "courted."

Why are sex pheromones important? Well, the answer isn't exactly romance, as the cartoon suggests. Sex pheromones are necessary for the survival of a species. They help males and females of the same species find each other so that they can reproduce.

Pheromones work just as well for insects that live in large groups. Pheromones broadcast information quickly, which can be very important when danger threatens. When a honeybee stings an enemy, the bee releases a pheromone that smells like bananas. The pheromone travels quickly through the air, sending an alarm message to other honeybees. When the other bees smell this alarm pheromone, they're attracted to it. They quickly gather and get ready to help defend the hive.

Sex pheromones are an important kind of chemical communication between males and females of the same species.

Checkpoint

1. Give an example of a chemical signal that a plant sends to another organism.
2. What are sex pheromones and why are they important in insect communication?
3. **Take Action!** By releasing a pheromone onto the ground, an ant can lay a scent trail that guides other ants to a food source. Use peppermint extract to make a scent trail. Then have your classmates try to follow the trail to the "food."

LESSON

1.3 Getting The Bugs Out

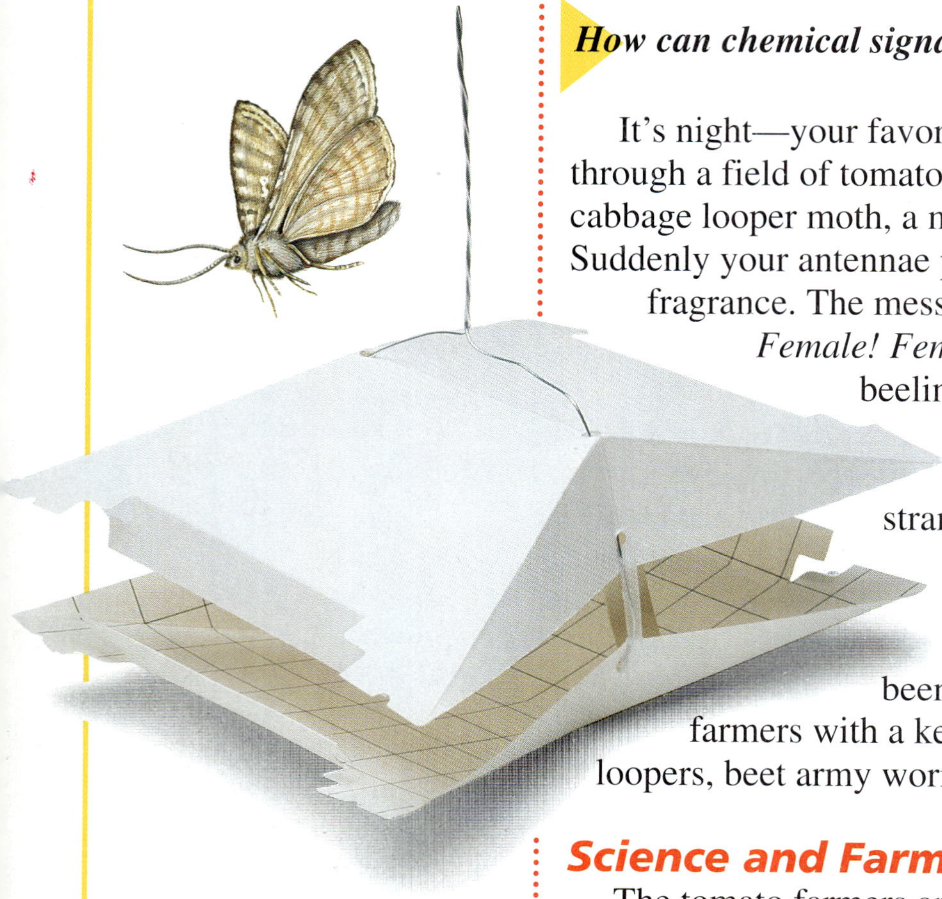

▲ *This pheromone trap is used to attract and trap many kinds of insect pests, including these cabbage looper moths.*

How can chemical signals help farmers?

It's night—your favorite time—and you're flying through a field of tomatoes. You're a fuzzy, brown cabbage looper moth, a male of your species. Suddenly your antennae pick up an *irresistible* fragrance. The message shrieks like a siren: *Female! Female! Female!* You make a beeline—oops, a mothline—toward the smell. Finally, you land in this strange-looking cardboard box. Hmmm. Other males are in here, but, hey, where's the female? You have been trapped by two tomato farmers with a keen interest in cabbage loopers, beet army worms, and other tomato pests.

Science and Farming

The tomato farmers are a husband-and-wife team who are working with scientists to find ways to reduce the amount of chemicals people put on crops. They catch male cabbage loopers by attracting them into traps that are baited with sex pheromones of female cabbage loopers.

The farmers count the male moths in the pheromone traps. The count helps farmers determine how many female moths are in the field laying eggs on the tomatoes. The eggs will hatch into larvae that will worm their way into the farmers' tomatoes.

If the numbers of trapped males are high, the farmers will apply insecticide to control the cabbage looper larvae. If the numbers are low, no insecticide is necessary.

Scientific knowledge about plant chemical signals can also be put to work in fields and gardens. For years, farmers and gardeners have used sprays containing pyrethrums as insecticides. The "Don't eat me!" message of pyrethrums keeps plant-eating insects away from crops.

With more research, scientists and farmers may find that the alarm chemical made by sagebrush plants can be useful. By spraying the chemical on fields, crops might begin to make the chemicals that protect them from plant-eating insects.

Today, more than ever before, scientists and farmers are working together to use what they know about plants and insects. Their work is helping find better ways to grow crops, such as the enormous wheat crop shown below. Thanks to their work, you can leave a grocery store with a shopping cart filled to the brim with tasty, nutritious foods!

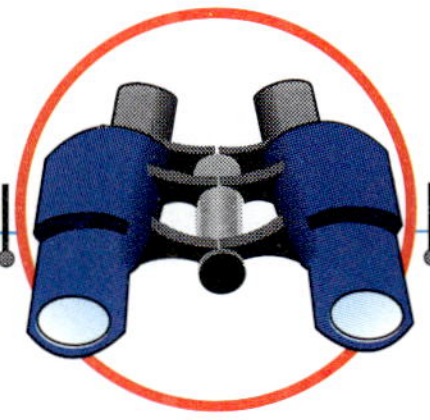

Into The Field

Can you design an insect trap that uses no chemicals?

Invent an insect trap that would be safe for the environment. Draw and describe your insect trap invention.

Farm machines called combines harvest a wheat crop.

The Age of DDT

Ever since farming began 8000 years ago, people have been trying to get rid of insects and other crop pests. Early methods of pest control were hard work. People tried trapping insects, swatting them, and scaring them away. In some parts of the world entire villages still tramp through the fields beating drums to scare off grasshoppers.

A history of pest control methods is shown in the timeline on the next few pages. Follow the dates and pictures as you read. In the 1800s, people began to use chemicals to kill insect pests. A chemical that kills pests that damage crops is a **pesticide** (pes′tə sīd). Back in the 1800s, most pesticides contained metals, such as arsenic, copper, and mercury. Some of these pesticides can be very poisonous. They often killed the plants as well as the insects!

In 1939, an accidental discovery made people in this century very dependent on pesticides. In that year, Swiss chemist Paul Müller discovered that a chemical called DDT killed moths that ate wool clothes. So he began to investigate DDT. Müller found that DDT was a powerful insecticide that killed many other kinds of insects.

Ancient times
Chrysanthemum petals are crushed to make the insecticide pyrethrum.

Late 200s
The Chinese begin to use yellow citrus killer-ants to control insect pests in their mandarin orange groves.

Mid 1800s
Vineyards in France are treated with carbon bisulfide to kill aphids that destroy grape vines.

1939
Dr. Paul Müller discovers that dichloro-diphenyl-trichloro-ethane, or DDT, can kill insects.

1942–1945
World War II soldiers use shaker-top cans to dust powdered DDT on their clothing.

1946
Airplanes are used to apply DDT over large areas of farmlands. Many airplanes were those used during World War II.

1948
Dr. Paul Müller wins a Nobel Prize for discovering that DDT could be used as a pesticide.

During World War II, DDT was used to kill disease-carrying insects such as head lice and mosquitoes. DDT saved thousands of soldiers who would have died of diseases, such as typhus and malaria, that these insects carry. Shortly after the war, Paul Müller won a Nobel Prize for his life-saving discovery.

In the 1950s, people turned their attention to a peacetime problem—how to produce food for the world's growing population. Farmers began to use DDT in their fields. At first, people were very excited about the new "wonder" chemical. Farmers discovered that DDT killed a wide variety of crop pests. But that's not all. DDT was also cheap and easy to use.

It wasn't long before farmers were using DDT on everything. Harvests were gigantic! It looked as if the problem of feeding the world was solved. But slowly people began to realize that DDT affected more than just insect pests. By the 1960s, people began to see that DDT was poisoning *everything*—the soil, rivers, lakes, wildlife, and people!

A New Day in Pest Control

The timeline on this page continues to 1972—an important date in the history of pest control. In 1972, a law was passed that made the use of DDT illegal in the United States. People's views about pesticides were changing. By that time a great deal of evidence showed that DDT was harming the environment. And the poisonous effects of DDT lasted a long time.

Studies also showed that DDT wasn't working anymore. Malaria was on the rise again. Why? Mosquitoes that carried malaria had developed resistance to DDT—the pesticide no longer had any effect on them.

Crop pests were also becoming resistant to DDT. The more the farmers applied DDT, the less it worked. Heavy use of DDT was causing more and more insects to become resistant to the chemical.

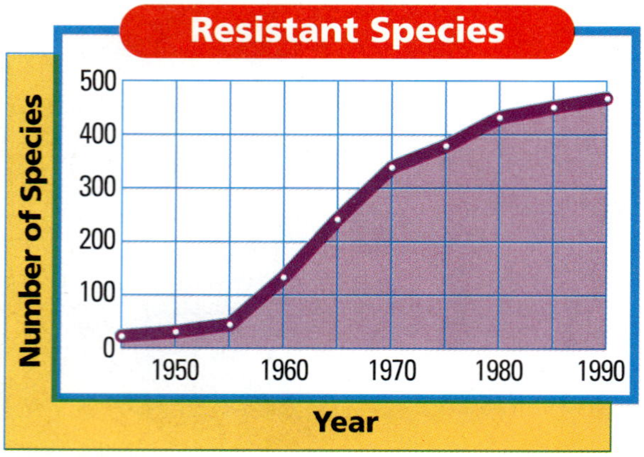

1955
The numbers of insects resistant to pesticides begin to increase sharply.

1960s
As the harmful effects of DDT become known, people begin to protest against the use of DDT.

Farmers were also discouraged to find that their fields had greater numbers of pests than ever before. What was going on? For one thing, DDT was also killing the natural enemies of insect pests. With their enemies wiped out, the populations of the pests could increase rapidly.

1972
The United States government bans most uses of DDT, including its use in farming.

But what has happened since 1972? Many chemical pesticides quickly replaced DDT, but they have the same problems. Many of these pesticides harm the environment. In fact, some of these chemicals are even more poisonous than DDT. And, as you can see in the graph, many kinds of insects have developed resistance to pesticides. Of these resistant species, many are disease-carrying insects and serious crop pests.

For more than 50 years, people have been waging all-out chemical warfare against insects. At one time, people believed that chemical pesticides would solve the problem of producing food for the world's growing population. But sometimes crop losses are just as high as they were before so many chemicals were used. By now, you probably wonder, "Who's winning this war?"

Solving the problems of pest control is an exciting challenge for scientists and farmers. How can they develop new chemicals and methods that do the job safely and inexpensively—and still grow enough food for people everywhere?

Many farmers are using smaller amounts of chemicals. Farmers are also applying chemicals to their fields less often. And some farmers, called organic farmers, don't use any chemicals at all. Farmers are also using biological pest control. A **biological control** is a method that uses natural enemies to control insects. As you read on, you will see that biological control is one way farmers can work *with*, not against, nature.

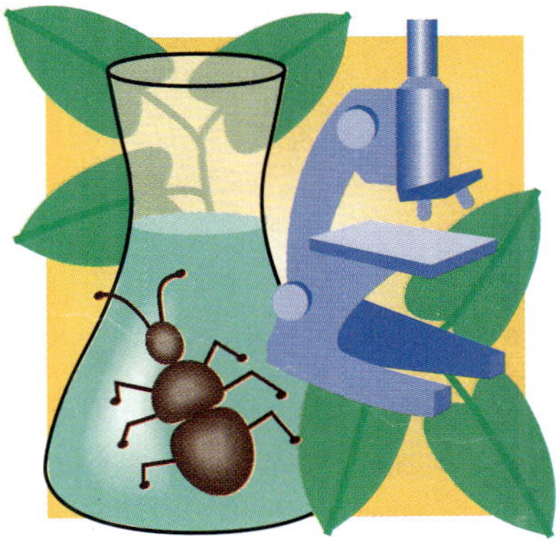

1976
Farmers begin to experiment with a new pest control method—integrated pest management.

1980s
Organically grown fruits and vegetables become available in many grocery stores and farm stands.

1990s
Researchers continue to develop new chemicals and methods to control crop pests.

Battle of the Bugs

Some insects can help solve pest problems on farms and in gardens.

You're a farmer with a big problem. Your farm has been invaded by pests. Grasshoppers are chewing your corn plants. Aphids are sucking sap from your pea plants. Weevils are gnawing holes in your pepper blossoms. Stalk borers are burrowing into the stems of your squash plants.

How are you going to solve your problem? Well, you just might want to bring in *more* insects! In nature, many kinds of insects kill other insects. In this way, the populations of certain kinds of insects are kept under control.

Today, many farmers and gardeners control pests the natural way. They are learning how to use helpful insects to control the number of harmful insects that attack their crops.

Praying mantis
Eats most kinds of insect pests, including weevils, grasshoppers, aphids, and stalk borers.

Ladybug
Eats mainly aphids, but also eats stalk borers, weevils, whiteflies, and Colorado potato beetles.

Braconid wasp

Lays eggs inside the larvae of cabbage looper moths, grain weevils, cabbage butterflies, and tomato hornworms.

Assassin bug

Attacks many kinds of leafhoppers, caterpillars, and aphids, as well as Colorado potato beetles.

Take a look at the insects shown here. They should be very welcome in gardens and on farms. Why? These helpful insects kill many kinds of crop-damaging pests. Here's how they do it.

Ladybugs and praying mantises are the lions and tigers of the insect world. These hungry hunters have chewing mouthparts, which they use to devour many kinds of garden and crop pests.

Assassin bugs are hunters, too. But these bugs have sucking mouthparts. An assassin bug holds its victim, or prey, with powerful front legs. Then the assassin bug uses its long, curved beak to pierce its prey and suck out its body fluids.

Some kinds of female wasps kill pests in another way. These wasps lay one or more eggs inside the body of an insect pest. When the egg hatches, the larva—or young—feeds on the insect's internal organs, killing it from the inside out.

Traps provide farmers with information about the kinds and numbers of pests in fields.

Natural enemies are used as biological controls of crop-damaging insects.

Pest-resistant varieties of crops are grown.

▲ *What are some advantages of using IPM?*

Integrated Pest Management

As you learned, many farmers are beginning to think that they can't solve pest problems with chemicals alone. Pesticides are getting expensive and they don't always work.

Some farmers are trying a different strategy—Integrated Pest Management, or IPM for short. It works like this. Farmers use a combination of ways to control insects. That's what the word *integrated* means. Some of the methods that are parts of IPM are shown in the diagram.

IPM is not easy. Farmers who use IPM must have a great deal of knowledge about both plants and pests. Farmers must also make careful, often difficult, choices. For example, when farmers find insect damage in their fields, they must decide what to do. Will biological control work or should a pesticide be used? Whenever possible, farmers avoid using chemicals to kill crop-damaging insects. They also try to avoid using weed killers, or herbicides (hėr′bə sīds). Instead, farmers control weeds by pulling them out.

Crops are planted in narrow rows to create shade that prevents weed growth.

Chemical pesticides are used only when other pest control methods fail.

In Indonesia, the government gave rice farmers a crash course in IPM. Farmers were spraying pesticides on their rice paddies up to eight times a season to kill pests called planthoppers. The more they sprayed, the more planthoppers there were. The pesticides were killing spiders that eat planthoppers!

So, the government banned many kinds of pesticides and taught the farmers ways to bring back the spiders. In three years, farmers were using 90 percent less pesticides. What about the spiders? They were back at work, eating planthoppers.

Checkpoint

1. How can farmers use insect pheromones?
2. List the benefits and dangers of DDT.
3. What is organic farming?
4. Give an example of biological control.
5. What is IPM?
6. **Take Action!** Read the warnings on flea powders to find out about the dangers of using these pesticides on pets.

DATAFILE

SOCIAL STUDIES

Evaluating IPM

In Indonesia, integrated pest management, or IPM, is being used to control planthoppers. Farmers are using far fewer pesticides to spray their rice fields. Instead, farmers are using spiders, which are natural enemies of planthoppers, as biological controls. This chart shows the results of Indonesia's IPM program.

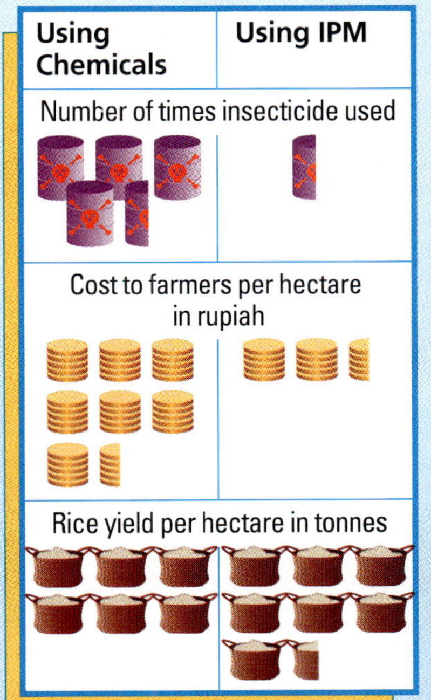

Pest Management

What Did You Find Out?

1. *How does the cost of IPM compare to the cost of using chemicals?*

2. *How do the two crop yields compare?*

3. *Which method of pest control got better results—IPM or spraying with pesticides?*

Activity

A Safe Way to Spray

There are many ways to control pests. In this activity you can make an insect spray that is safe for you to use.

Picture A

Picture B

Picture C

Gather These Materials

cover goggles
1 garlic bulb
1 small onion
plastic knife
cutting board
2 mixing bowls
large measuring spoon
cayenne pepper
liter measure
water
mixing spoon
liquid soap
cheesecloth
2 spray bottles
large jar with lid
2 mature plants, same type and size

Follow This Procedure

1. Make a chart like the one on the next page. Record your observations in the chart.
2. Put on your cover goggles.
3. Carefully chop the garlic bulb and onion into tiny pieces. Then place them in a mixing bowl. (Picture A) Add one spoonful of cayenne pepper to the bowl.
4. Add 1 liter of water and stir. Let the mixture sit for one hour. Then stir in one spoonful of liquid soap. (Picture B)
5. Pour the mixture through a few layers of cheesecloth into another mixing bowl. Then pour some of the liquid into a spray bottle. Pour the rest into a jar. Close the jar tightly, and put it in the refrigerator. It will keep for up to one week.

6 Fill the other spray bottle with water.

7 Put both plants outside. Once each day, lightly spray one plant with your insect spray. Spray the other plant lightly with water. (Picture C)

***Predict:* How do you think the insect spray will affect the plant?**

8 Observe the plants over the next few days. Water the plants as needed. Watch for insects. Record any differences between the plants in the chart.

Record Your Results

Observations		
Date	Sprayed with insect spray	Sprayed with water

State Your Conclusions

1. What are some of the advantages to using this kind of insect spray?

2. Did you notice any differences between the plant treated with the insect spray and the plant that was not treated?

Let's Experiment

Do you think your spray would keep animals away from vegetable plants in a garden? Use what you know about scientific methods to find out.

Chapter Review

Reviewing Words and Concepts

Write the letter of the word or phrase that best completes each sentence.

1. A ____ is a chemical messenger with a precise set of instructions.
2. Weed-killing chemicals are called ____.
3. *Rhizobium* ____ live in nodules on the roots of legumes such as beans, peas, and alfalfa.
4. A ____ communicates messages between two or more members of the same species.
5. A chemical that kills pests that damage crops is a ____.
6. Pests that no longer are killed by a chemical that once killed them are ____ to the chemical.
7. A ____ is a method that uses natural enemies to control insects.
8. In the 1950s, ____ was a common pesticide used on crops.
9. The praying mantis, ladybug, and assassin bug eat many kinds of ____.
10. Using a combination of ways to control insect pests is called ____.

a. resistant
b. biological control
c. pheromone
d. herbicides
e. bacteria
f. IPM
g. pesticide
h. pests
i. DDT
j. hormone

Connecting Ideas

1. Copy the concept map. Use the terms at the right to complete the map about chemical signals.

pheromones **hormones**
growth **reproduction**

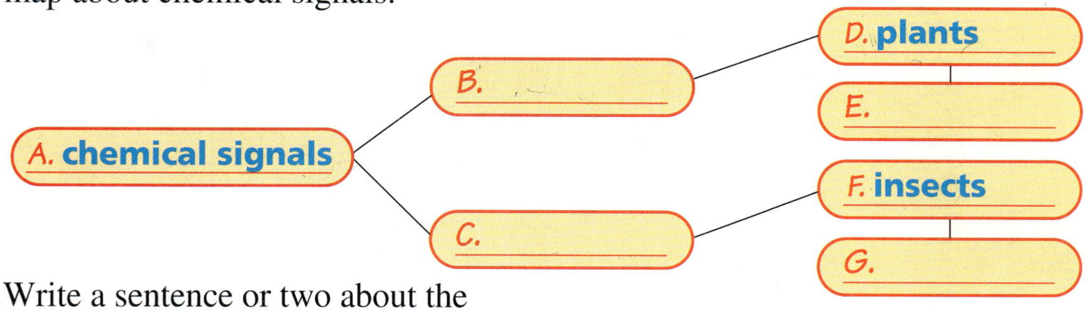

2. Write a sentence or two about the ideas shown in the concept map.

Interpreting What You Learned

1. What is the "plant talk" hypothesis?
2. What effects do environmental signals have on the growth and development of a plant?
3. How are legumes and *Rhizobium* bacteria helpful to each other?
4. How are sex pheromones important for the survival of a species?
5. Identify a positive and a negative effect of DDT.
6. List three pest control methods used in integrated pest management.

Performance Assessment

How can you model how plants repel insect pests?

Materials
- perfume
- garlic
- onion
- orange slice
- tape
- green construction paper
- small paper bag
- dropper
- scissors
- pencil

Collecting Data

1. Draw four identical leaf shapes on a piece of green construction paper. Cut out the leaf shapes.
2. Place a drop of a perfume on one of the paper leaves. Gently blow on the drop until it is dry.
3. Rub a piece of garlic over the surface of another paper leaf. Gently blow on the surface of the leaf to make sure it is dry.
4. Rub a piece of onion over the surface of another paper leaf. Gently blow on the surface of the leaf to make sure it is dry.
5. Rub an orange slice over the surface of the last paper leaf. Gently blow on the surface of the leaf to make sure it is dry.
6. Place the cutout leaves in a paper bag. Pretend you are an insect that is searching for leaves to eat. Remove one leaf from the paper bag. Smell the leaf. Decide whether the leaf is making chemicals that cause it to taste bad. Describe what you smell and tell whether you would eat the leaf. Tape the leaf on your paper next to your description.
7. Repeat step 6 using the other three leaves. Describe what you smell and tell whether you would eat each leaf. Tape the leaves on your paper next to the descriptions you write.

Analyzing Data

If you were an insect, which of the leaves would you be most likely to eat? Which of the leaves would you be least likely to eat?

CHAPTER

2 Growing Plants

Discover Activity

What's the best way to grow peas?

Select a packet of pea seeds. Think about what conditions might affect the growth of peas. Design an experiment to find out which conditions are best for growing pea plants. Observe your plants every day for three weeks. Record your observations.

For Discussion

1. *What conditions did you test and why?*
2. *What conditions are best for growing pea plants?*

LESSON

2.1 Breaking New Ground

What makes plants grow?

Maybe you'd like to ask this young gardener, "What's your secret?" Well, here's what he'd tell you, "Secrets? Forget it! Growing sweet peppers isn't a big mystery. Pepper plants need six things. Supply them with these needs and you'll harvest a basketful of sweet, red peppers!"

Checklist for Growing

As you observed in the Discover Activity, plants grow and develop. Think about your pea plants. They grew from seeds into plants with slender stems, and leaves with curly threads called tendrils. If the pea plants were growing in a garden, they would have formed pods containing seeds. What does a pea plant need to grow? It must have those six things that the pepper grower talked about. To grow, plants need light, water, nutrients, the right temperature, gases in the air, and space to grow.

To produce these tasty peppers, the pepper plants need air, light, warm temperatures and plenty of space. They also need to be watered and fertilized often.

Hydroponics: All A Plant Needs

On hydroponic farms, crops grow indoors in a nutrient-rich liquid.

Plants that grow on farms get most of the things they need from nature. But sometimes nature is undependable. Bad weather may prevent crop plants from getting the sunlight, water, or warmth they need to grow well.

Now scientists and farmers are experimenting with hydroponics, a new kind of agriculture. **Hydroponics** is the science of growing plants in a nutrient-rich liquid instead of in soil. On hydroponic farms, crops such as lettuce and spinach, are raised indoors in factory-like buildings. The plants are supplied with the things they need to grow well—air, light, nutrients, water, and proper temperature.

Temperature

Air temperature is carefully controlled so that plants get the warm temperatures they need. Electricity and fuels are used to heat the building.

Water

A system of pipes and sprinklers supplies plants with water. The roots of the crops hang down into a solution of water and nutrients.

Nutrients from Soil

Unlike plants in hydroponic farms, most plants grow in soil. Soil is made mostly of tiny bits of crushed rocks. As plants and animals die and decay, their remains mix with the soil, too. The plant-and-animal part of soil is called humus (hyü′məs). This wonderful mix of rock and humus contains nutrients—chemicals plants need to grow. Soil is like a bank. Plants draw nutrient "riches" from the "soil bank" just as people take money from a bank.

With your finger, trace the path of the nutrient cycle. Explain how humus is part of this cycle.

Some of the nutrients in soil are minerals such as phosphorus, potassium, sulfur, calcium, and magnesium. Another soil nutrient is nitrogen, the hardest nutrient for plants to get! Three-fourths of the air is nitrogen gas. But plants can't use this nitrogen. They must take in nitrogen from the soil. Some soil bacteria change nitrogen gas into nitrogen compounds that plants can use. Nitrogen compounds are also formed when dead plants decay. Plants take in and use these compounds.

The diagram shows how plants use nutrients over and over again in a cycle. In nature, plants take nutrients from the soil. When plants die and decay, the nutrients are returned to soil as humus.

Now look at the diagram on this page. It shows that some nutrients on farmland aren't cycled, as they are in nature. When farmers harvest their crops, they take away parts of the plants that would die and become humus. It's a lot like taking money out of the bank but not putting money back. Then you have to borrow money.

When farmers take riches from the soil bank, they must borrow nutrients from fertilizers. A **fertilizer** is a substance, either chemical or natural, that adds nutrients to soil. Fertilizers are like deposits that replace money taken out of a bank. They allow farmers to keep growing crops on the land after nutrients have been removed.

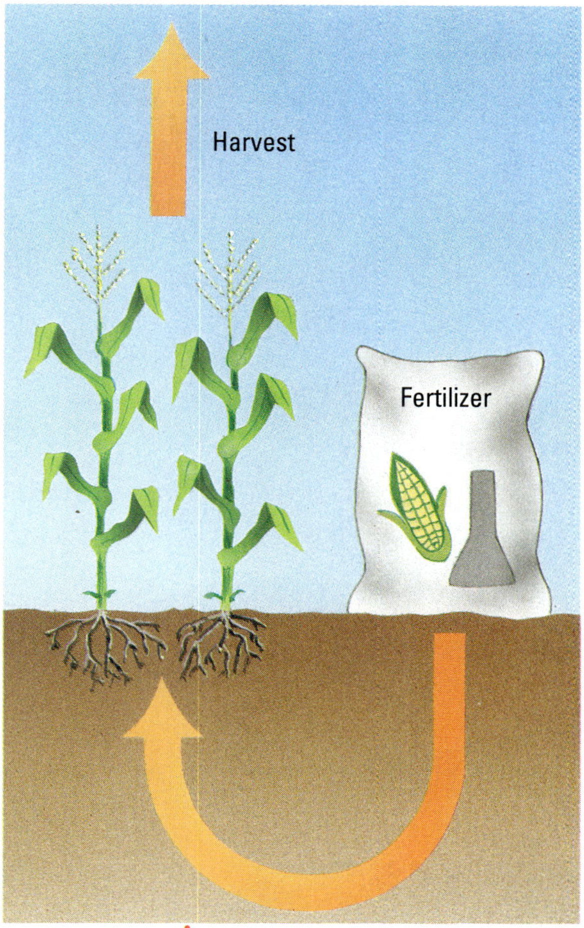

The diagram shows how the nutrient cycle can be broken on farmland. How do chemical fertilizers help complete the cycle?

Chemical fertilizers come in different forms, such as the particles you see here. Natural fertilizers include manure from cattle and other livestock. Legumes, which work with *Rhizobium* bacteria to make nitrogen, are another natural fertilizer. Sometimes farmers plant alfalfa and then plow the crop into the soil. Crops that add nitrogen to the soil in this way are called green manure.

Checkpoint

1. What things do peppers need to grow?
2. List the benefits of hydroponic farms.
3. How does farming break nutrient cycling?
4. **Take Action!** Examine a soil sample. Separate the humus from the rocks.

Activity

Doing Away With Dirt

Can plants really grow without soil? Try this activity to see how plants grow in a hydroponic solution.

Picture A

Picture B

Picture C

Gather These Materials

cover goggles
toothpicks
2 scallions
2 narrow jars
water
marker
hydroponic solution
scissors
black construction paper
tape

Follow This Procedure

1. Make a chart like the one on the next page. Record your observations in the chart.
2. Put on your cover goggles.
3. Poke 2-3 toothpicks through the leaves of a scallion, about halfway down from the top. Then, place the scallion in a jar so the toothpicks rest on the jar's rim. The bottom of the scallion should not touch the bottom of the jar. (Picture A)
4. Pour water into the jar until it reaches just above the scallion's roots. Mark the level of the water on the jar with a marker. (Picture B)
5. Prepare another scallion as in step 3. Place it in a jar, but add hydroponic solution instead of water.
6. Wrap a piece of black construction paper around each jar, and fasten the paper with tape. The paper will keep out light, preventing algae growth in the jars. (Picture C) Place the jars in a warm, sunny place.

Record Your Results

Growth rate observations		
Date	Scallion in water	Scallion in solution

***Predict:* Will the 2 scallions grow at the same rate?**

7. Check the jars every day, and add water to keep the liquid level in each jar constant. Do not add hydroponic solution to either jar.

8. Observe any changes in the scallions. Watch for root growth and leaf sprouting. Record your observations in the chart.

State Your Conclusions

1. Compare the growth of the scallions in the 2 jars.
2. What other plants do you think could be grown this way?

Let's Experiment

Do you think you could root cuttings in a hydroponic solution? Use what you know about scientific methods to find out.

LESSON

2.2 Bigger and Better

Heads of wheat form at the top of each wheat stem. Each head is made up of clusters of flowers.

How do plants form seeds?

If you break open a sweet pepper, you'll see dozens of tiny seeds that can grow into new pepper plants. Many plants reproduce by forming seeds. When planted, a seed develops into a new plant of the same kind. How do seeds form? The seeds of fruits and vegetables develop in flowers.

How Seeds Form

How would you describe a flower? Would you use words like *beautiful*, *colorful*, *fragrant*, *delicate*? You probably wouldn't say *factory*, but that's a good word to describe a flower. Flowers are like factories that manufacture seeds.

Take a close look at this wheat flower as you read about the mass production of seeds. Most flowers have both the male and female machinery needed to produce seeds. The male parts are long, skinny stalks called stamens. The tips of the stamens contain millions of tiny pollen grains. Pollen contains the male sex cells, or sperm.

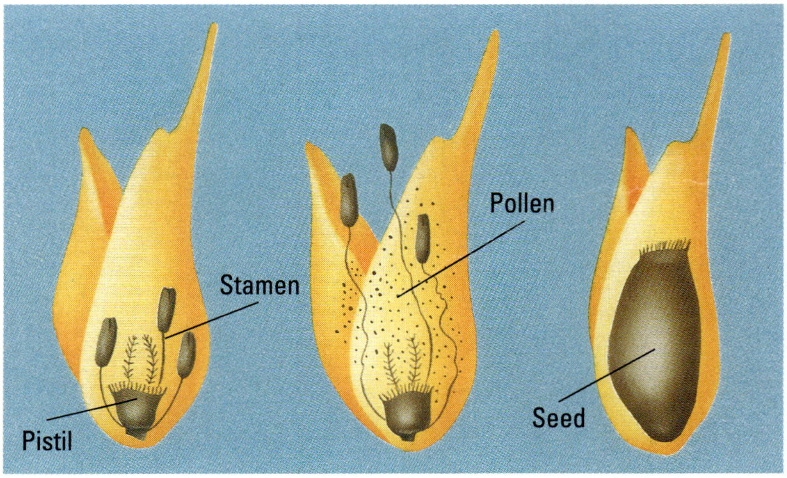

How a wheat seed forms

Many of the vegetables you eat are actually fruits. Which one of these vegetables is a flower?

The stamens surround the pistil, or female part of the flower. The top part of the pistil is usually sticky. The bottom part, or ovary, contains one or more ovules. Each ovule holds a female sex cell, or egg. When an egg and a sperm unite, a seed is formed.

But before a seed is formed, pollen must get to the pistil. The moving of pollen from a stamen to the sticky part of the pistil is called **pollination** (pol′ə nā′shən). This wheat flower pollinates itself. But sometimes pollen from one flower is carried by wind to another flower. Insects help pollinate other kinds of flowers. How? Pollen gets stuck to an insect's body when it lands on a flower. When the insect visits another flower, the pollen can rub off onto the sticky part of the pistil.

After pollination, a tube grows down from a pollen grain stuck to the pistil. A sperm cell from the pollen grain travels down the tube and joins with an egg inside an ovule. Then the egg cell develops into a seed—a grain of wheat! In other kinds of flowers, the ovary becomes a thick fruit that covers and protects the seed or seeds. Most of the vegetables in the picture are fruits.

Hybrids of Corn

People can control pollination to produce new varieties of plants.

What pollinated the plant that produced this tasty ear of corn—wind, an insect? It may have been a scientist!

The varieties of corn shown have different characteristics, or traits. Scientists can produce certain traits in corn and other plants by growing hybrids. A **hybrid** is an offspring of parents with different traits. To produce hybrids, scientists cross, or mate, two varieties of plants by controlling pollination. The hybrid produced by a cross has desirable traits from both parents.

Making hybrids
Scientists collect pollen from the tassel of one plant and transfer the pollen to the silk of the second plant. The second plant produces seeds, which grow into hybrid plants.

Hybrids of corn
Scientists have produced many new varieties of corn, especially hybrids of white and yellow sweet corn. Hybrids of popping corn make wholesome snacks. Hybrids of decorative corn—with red, orange, yellow, and blue kernels—make festive decorations.

Hybrid seeds
Many kinds of hybrid seeds are available, and new hybrids are developed every year.

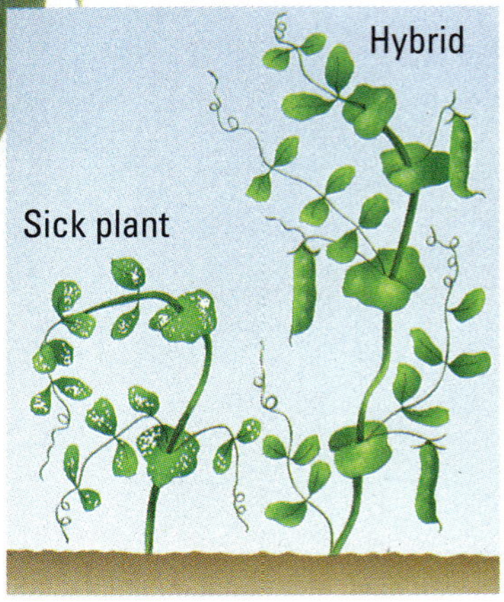

These peas were picked from a hybrid plant that is resistant to downy mildew. The diagram shows the effects of this disease.

New Foods From Hybrids

Did you ever wonder why scientists spend so much time developing new varieties of plants? Growing hybrids solves a lot of problems for gardeners and farmers. Thanks to hybrids, farmers can grow bigger, healthier crops. That's very important in the world today, because the growing population needs a plentiful supply of nutritious foods. Growing hybrids can also help protect the environment. How? Some hybrids don't get certain diseases. Other hybrids are not as likely to be eaten by certain crop pests. When farmers plant these hybrids, they don't need to use as many chemical pesticides on their crops.

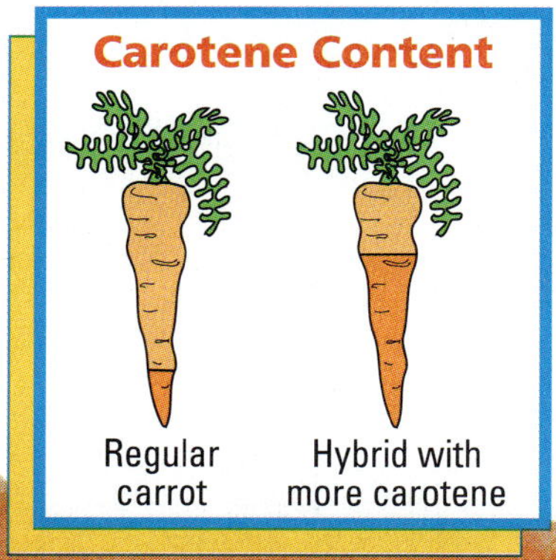

The diagram shows why this hybrid carrot is a more nourishing snack than a regular carrot.

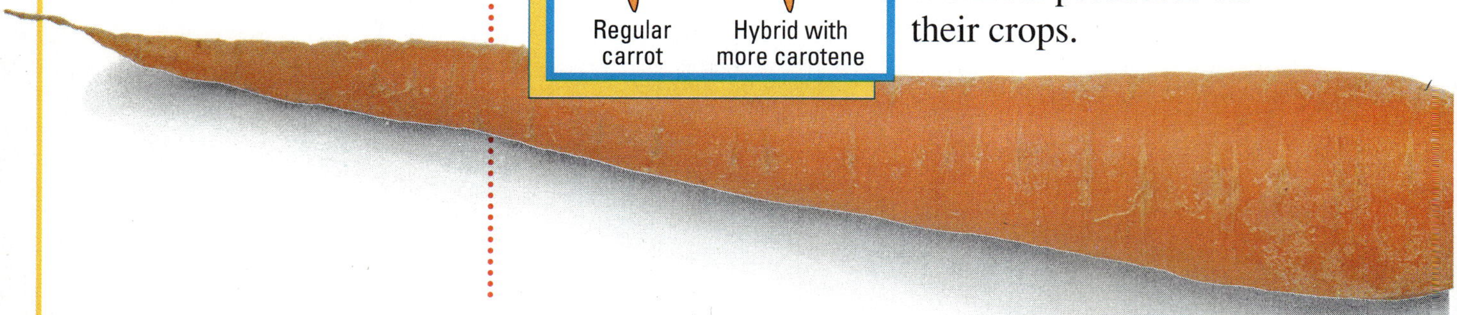

Growing Regions

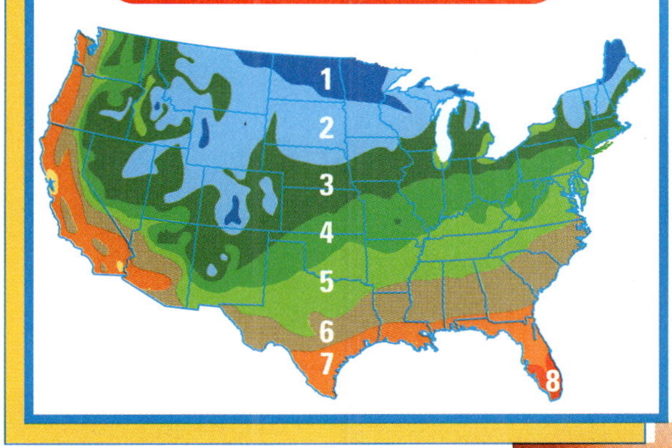

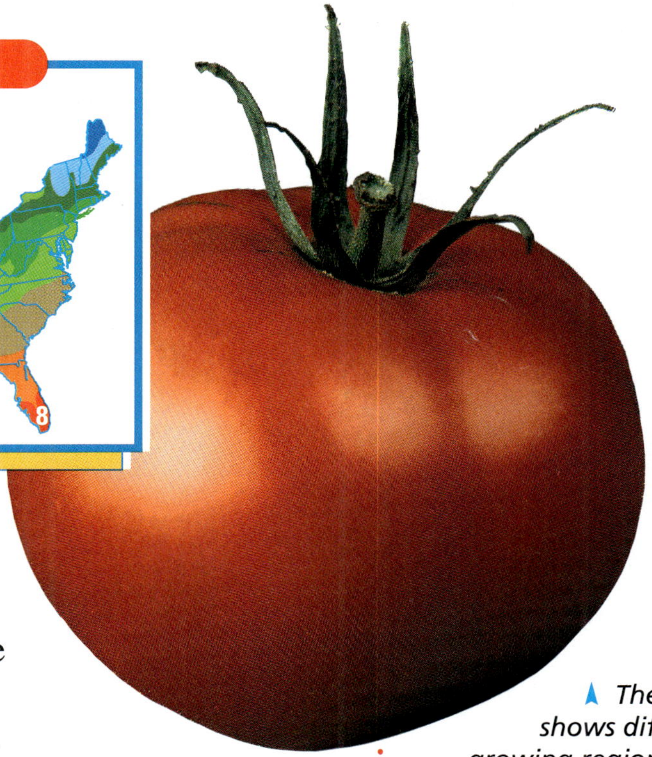

The vegetables on these pages are grown from hybrid plants. The peas come from pea plants that are resistant to a disease—called downy mildew—caused by a fungus. By growing this hybrid, a farmer doesn't have to use a fungicide (fun′jə sīd), or chemical that kills the fungus. These carrots not only look good—they're good for you! They're a hybrid that contains more carotene, a source of vitamin A. Farmers will tell you that tomato plants are *slow* growers. They need long summers to bear fruit. But this tomato was picked from a hybrid plant that produces tomatoes in a shorter period of time. It's grown by farmers and gardeners who live in the northern parts of North America, where summers are short. On the map, these northern regions are shown in blue.

▲ *The map shows different growing regions. The numbers stand for average temperatures in these regions. The smaller the number, the colder the region and the shorter the growing season.*

Checkpoint

1. What is pollination?
2. How do scientists produce hybrid plants?
3. Describe a way in which a hybrid can improve a food crop.
4. **Take Action!** Think of new hybrids that would improve your favorite vegetables. Make a poster of your new foods.

Into The Field

How many plant hybrids have you eaten?

Look for plant hybrids in seed catalogs or on seed packages. Record all the different plant hybrids you have eaten.

Activity

Prize Tomatoes

When it's time to plant, gardeners can choose from a wide variety of tomato hybrids. Try this activity to see how the hybrids compare.

Picture A

Picture B

Picture C

Gather These Materials

- several packages of different kinds of tomato seeds
- large flower pots
- soil
- fertilizer
- water
- garden stakes
- string or cloth strips

Follow This Procedure

1. Make a chart like the one on the next page. Record your observations in the chart.
2. Work as a member of a team for this activity.
3. Fill a pot with loosely packed soil. (Picture A) Choose a package of hybrid tomato seeds. Plant 4 seeds in the pot. Follow the growing instructions on the seed package. (Picture B)
4. Place the pot in a warm, sunny place. Add fertilizer and water as needed. (Picture C) Make sure that the growing conditions are the same for each team's pot of tomato plants.

Predict: ***How will the tomato plants of each team differ?***

5. Check your plant every day. When the seeds have sprouted, remove all but the largest sprout. When the plant is about 15 cm tall, use cloth strips to tie a stake to the plant to help it stand up.

Record Your Results

Observations			
Date	Growth rate	Flowering	Fruit production

6 Care for your tomato plant over a period of several months. Observe how quickly your plant grows, when it flowers, and when it produces fruit. Note how many tomatoes your plant produces. Record this information. Compare your data with those of other teams.

7 Compare the package description of your hybrid to your actual results. Then, compare your tomato plant to the plants of other teams.

State Your Conclusions

1. How did the growth of the various plants compare to their descriptions on the packages of seeds?
2. How did the various tomato plants and tomatoes compare?
3. Which hybrid would you choose for your garden? Why?

Let's Experiment

How do you think you could improve the growing conditions to make the best-growing tomatoes grow even better? Use what you know about scientific methods to find out.

LESSON

2.3 A New Crop of Crops

How can a plant be part firefly?

Here's a riddle for you to solve. What has green leaves and glows like a firefly? Did you guess a firefly plant? Well, firefly plants don't exist. Or do they? In a famous experiment shown here, scientists made a kind of "hybrid" using a firefly and a plant. A new technology makes it possible to cross two very different kinds of living things. What is this new technology? Before you learn about it, you need to learn about genes.

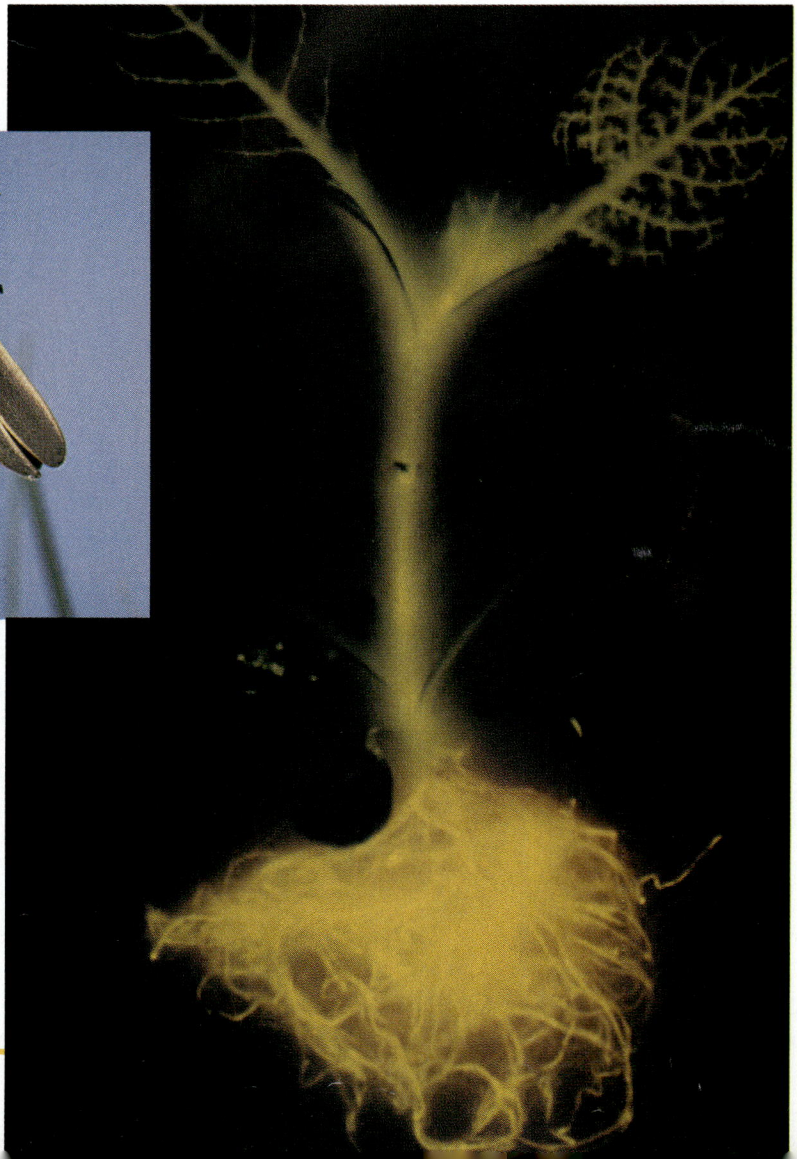

In the laboratory, scientists were able to make this plant glow like a firefly.

Working With Genes

Take another look at the firefly. It's easy to see that this organism has a set of characteristics, or traits—six legs, hard outer wings, yellow stripes, saw-shaped antennae. The organ that produces a firefly's glow is another trait. Each kind of organism has its own traits.

The cells of organisms contain information that controls traits. For example, a firefly's cells contain the information about the organ that makes a firefly glow. A wheat plant's cells contain information about the color and shape of wheat flowers. Where is this information? The nucleus of each cell contains threadlike strands called chromosomes (krō′mə sōmz). Each chromosome is divided into sections called genes. A **gene** is a unit of information that controls a trait.

Each kind of living thing has a set of genes that controls its traits. Genes that control a firefly's traits are different from genes that control a plant's traits. And the genes of different organisms do not mix. Why? In nature fireflies mate with fireflies, cats mate with cats, wheat plants pollinate wheat plants. You won't find a firefly plant in nature. So how did scientists make one in a laboratory?

As scientists learned about genes, they wondered whether a gene from one organism could be put into another organism. They developed a technology to do just that. **Genetic engineering** is a way to change the traits of an organism by changing its genes. Using genetic engineering, scientists removed a gene from a firefly and added it to a plant. The plant glowed like a firefly!

Scientists are using genetic engineering to produce new varieties of crop plants. These new plants can help farmers in the same ways hybrids do. For example, scientists might be able to give crop plants new genes that would protect the plants from insect pests. Then farmers wouldn't have to use as many pesticides on their crops.

DATAFILE

SCIENCE

Locating Your Genes

Like all humans, you have 23 pairs of chromosomes. Each chromosome is made up of thousands of sections called genes. Scientists are learning to locate specific genes on each human chromosome. For example, scientists have found the locations of genes for certain inherited diseases.

This chart shows the approximate number of genes in a human. The chart also shows the number of human genes whose exact location is known.

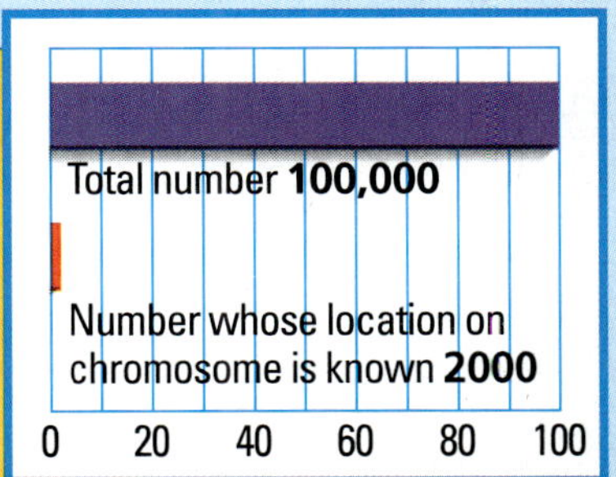

Human Genes (thousands)

What Did You Find Out?

1. *About how many genes do you have?*

2. *About how many of these genes have been located? How many must still be located?*

3. *What would be the advantage of locating the gene for a specific disease?*

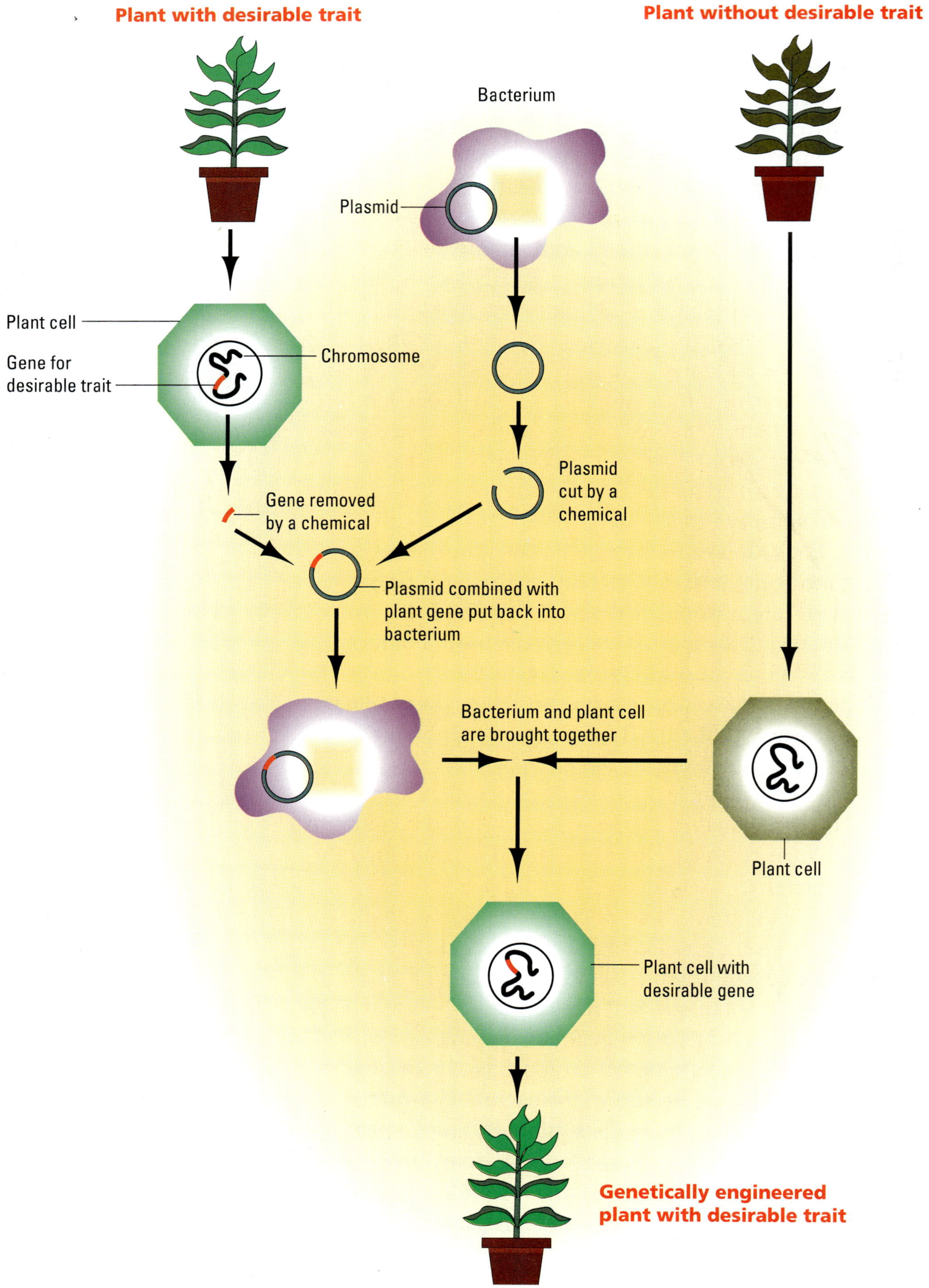
Plant with desirable trait
Plant without desirable trait
Bacterium
Plasmid
Plant cell
Gene for desirable trait
Chromosome
Plasmid cut by a chemical
Gene removed by a chemical
Plasmid combined with plant gene put back into bacterium
Bacterium and plant cell are brought together
Plant cell
Plant cell with desirable gene
Genetically engineered plant with desirable trait

Genetic Engineering

How does genetic engineering work? Scientists use several different methods. One way is to use a carrier—an organism that carries the gene from one organism to another. This method is shown in the diagram. Touch each step as you follow the procedure of transferring a gene from one kind of plant to another kind of plant.

First, scientists choose two plants. One of the plants has a desirable trait, such as resistance to a particular disease. The second plant does not have this desirable trait. Next, scientists take a cell from the plant with the desirable trait, and remove the gene that controls the trait. How do they do it? Scientists use a special chemical to cut the gene out of the chromosome.

Then the gene is put inside a carrier organism—a bacterium. Remember, bacteria are simple, one-celled organisms. Many kinds of bacteria have a ringlike structure called a plasmid. Scientists remove the plasmid from the bacterium and cut it so that it has two "sticky" ends. Then scientists "glue" the plant gene to the sticky ends of the plasmid. Finally, the plasmid is put back into the bacterium. Now the bacterium is ready to carry the new gene to the second plant.

In the last steps, scientists take a cell from the plant that doesn't have the trait. They put this cell into contact with the bacterium. The bacterium infects the plant cell with a disease. During the infection, the new gene becomes part of the chromosome in the plant cell.

But one more step remains. The plant cell—with its new gene—must grow into a new plant, with roots, stems, and leaves. Scientists place the cell in a liquid that contains nutrients and plant hormones. The plant cell grows and develops into a young plant. All the cells in the new plant have the gene that controls the new trait. So, the genetically engineered plant has resistance to disease!

This scientist is growing new plants from plant cells.

These farmers sell their vegetables at an open air market in India.

Designer Foods

How might plants with new genes help farmers, like those in this picture, grow food? Well, think about what plants need to grow. Remember that plants need nutrients. Ask farmers to name the secret to growing large crops, and they'll say, "Nitrogen!" Ask farmers to name a nutrient that is often missing in soil, and they'll say, "Nitrogen!" To make up for this shortage, many farmers use chemical fertilizers that contain nitrogen.

But chemical fertilizers are getting more expensive. Also, they can be harmful to the environment. When rain falls, fertilizers wash from the soil into rivers and lakes. The nutrients in fertilizers cause water pollution. To solve these problems, scientists are trying to develop new kinds of plants that don't need to be fertilized. Instead, the plants would make their own natural fertilizers—nitrogen compounds. Remember that legumes make nitrogen compounds, with the help of *Rhizobium* bacteria. Now scientists are planning different ways to get other crop plants—corn, wheat, and rice—to make nitrogen compounds, too.

One kind of experiment is shown in the diagram. You might call it the "fix it yourself" experiment. Why? It's working with the group of genes that control *Rhizobium's* ability to "fix" nitrogen, or change nitrogen gas into nitrogen compounds. Using the methods you just learned about, scientists plan to remove the genes from *Rhizobium* bacteria.

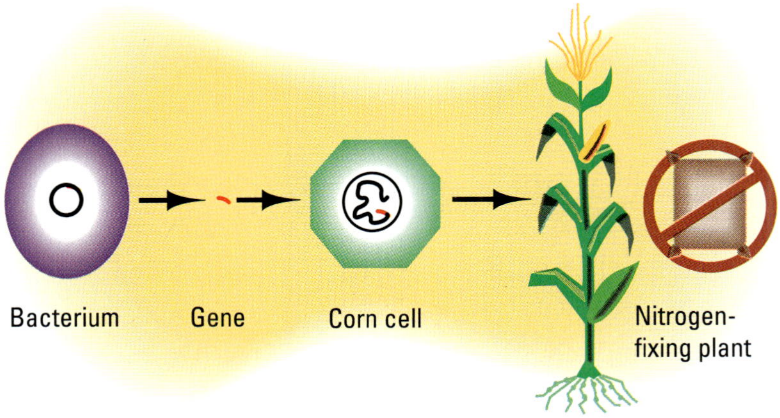

A nitrogen-fixing corn plant might be produced by genetic engineering. As you can see, this plant would not need chemical fertilizer.

Next, the scientists would try to put the genes into cells of corn, wheat, or rice plants. The scientists hope that the genetically engineered plants would be able to do what *Rhizobium* does—make nitrogen compounds. If these genetic engineering experiments work, just think about what it would mean. All the corn fields and wheat fields in the world might not need to be fertilized for nitrogen!

Checkpoint

1. What does genetic engineering allow scientists to do?
2. How do scientists use bacteria in genetic engineering experiments?
3. How are scientists using genetic engineering to develop new crop plants?
4. **Take Action!** Make a diagram that shows how a gene from a soybean plant might be put inside a corn plant to produce a corn plant with nodules.

Chapter Review

Reviewing Words and Concepts

Write the letter of the word or phrase that best completes each sentence.

1. Growing plants in liquid rather than in soil is called ____.
2. The part of soil that contains decayed animals and plants is called ____.
3. A ____ is a substance farmers add to soil to increase the nutrients in it.
4. When pollen moves from a stamen to a pistil, ____ occurs.
5. A ____ is an offspring of parents that have different traits.
6. A ____ is a unit of information that controls a trait.
7. Traits of an organism can be artificially changed through ____.
8. Genetic information can be transferred from one organism to another using a carrier such as a ____.
9. Many bacteria contain a ringlike structure called a ____.
10. To ____, a plant must be able to change nitrogen gas into nitrogen compounds.

a. gene
b. bacterium
c. pollination
d. plasmid
e. genetic engineering
f. hydroponics
g. fix nitrogen
h. hybrid
i. humus
j. fertilizer

Connecting Ideas

1. Copy the concept map. Use the terms at the right to complete the map about how pollination occurs.

pistil **pollen** **pollination**

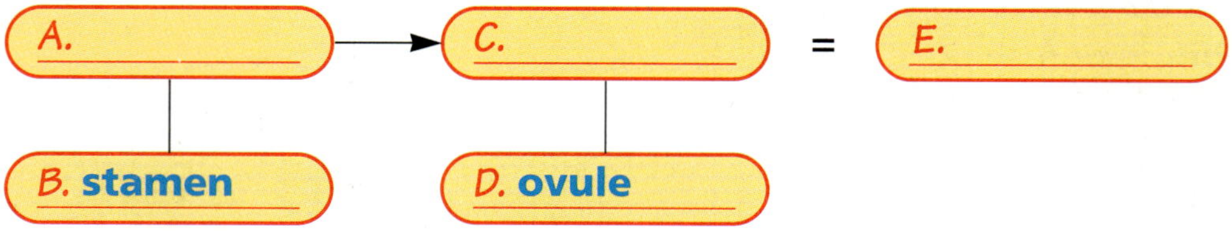

2. Write a sentence or two about the ideas shown in the concept map.

Interpreting What You Learned

1. Identify six things all plants need in order to grow and develop.
2. Explain how bacteria in the soil can help plants grow.
3. Describe the role insects play in pollination.
4. Explain how growing hybrids can help protect the environment.

Performance Assessment

How can you design a hybrid?

Materials
- 2 flower seed packages
- 2 vegetable or fruit seed packages
- crayons or markers
- 2 sheets of construction paper

Collecting Data

1. Obtain two different flower seed packages. Each package should contain information about traits such as plant height, germination time, color of flower, time until bloom, and so on. List the traits that both packages describe. Record these traits for each plant as they are listed on the packages.
2. Pretend that you have created a hybrid flowering plant that combines the traits of these two flowering plants. Decide how each of the old traits would change to form new traits in the hybrid. Record the new traits of the hybrid.
3. Draw a picture of the two parent flowering plants and the new hybrid flowering plant. Include a scale that shows how their heights compare.
4. Obtain two different vegetable or fruit seed packages. Each package should contain information about traits such as plant height, germination time, size and color of vegetable or fruit, time until harvest, and so on. List the traits that both packages describe. Record these traits for each vegetable or fruit as they are listed on the packages.
5. Pretend that you have created a hybrid vegetable or fruit that combines the traits of these two vegetables or fruits. Decide how each of the old traits would change to form new traits in the hybrid. Record the new traits of the hybrid.
6. Draw a picture of the two parent vegetables or fruits and the new hybrid vegetable or fruit.

Analyzing Data

Describe how either of the hybrids is more desirable than its parent plants.

CHAPTER 3

Gardening Plants

I wonder if any worms are in here!

Discover Activity

How can you grow an indoor mini-garden?

Think about what you learned about growing plants. Use this information to plan an indoor mini-garden. Choose the seeds you want to plant and the growing conditions for them. Observe your garden for about three weeks.

For Discussion

1. *How well did your garden grow?*
2. *What choices did you make in planning and caring for your garden?*

LESSON

3.1 Growing A Garden

How would you grow a vegetable garden?

You've learned a lot about plants and what they need to grow. When you did the Discover Activity, you used what you know to grow an indoor garden.

Now think about how you would plant a garden outdoors. Where would you begin? A good place to start would be to collect the items you see here. You'll also need a garden book. Why? It will tell you what your plants need to grow. For example, each kind of plant needs a certain kind of soil, a certain amount of sunlight, and a certain kind of fertilizer. If you don't know what your plants need, you won't be harvesting many vegetables!

Pretend you are going to plant and care for a garden. Then play the garden game on the following pages. Pick the crop that you'd most like to grow. As you read, use the boxes to make decisions about how to grow your crop. Each box has four choices. Pick the choice you think is best for *your* crop. When you're finished, see whether your choices helped your crop grow.

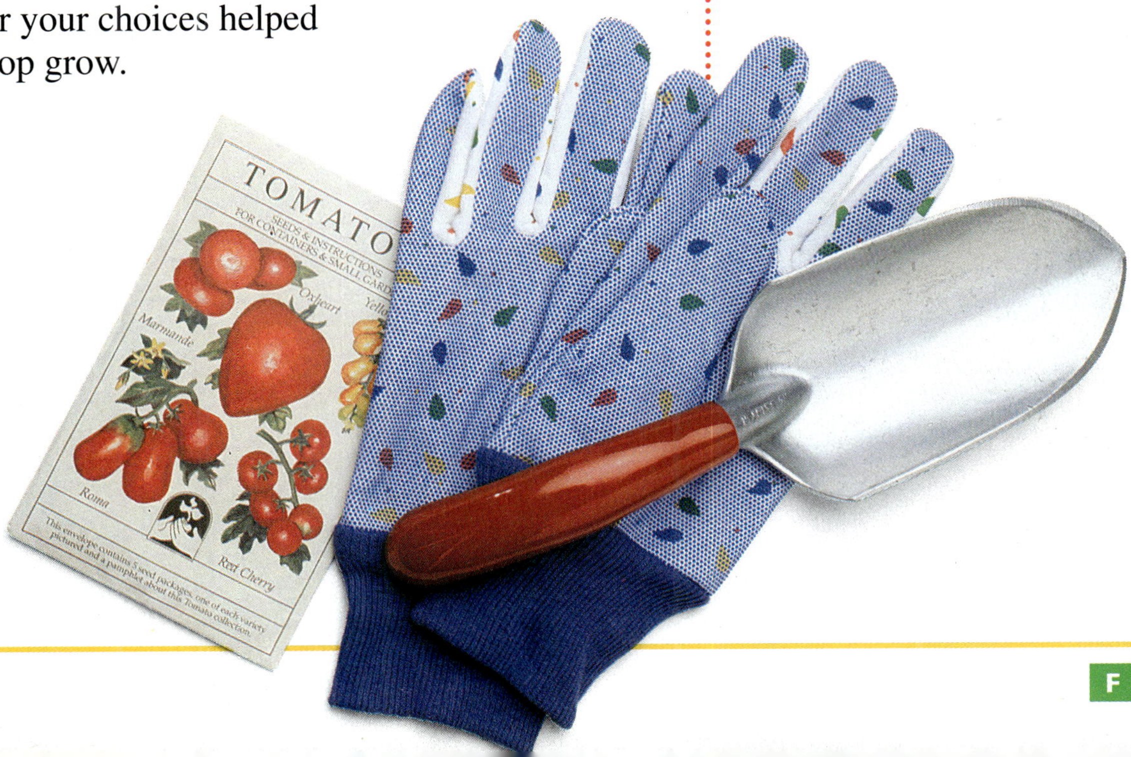

Garden gloves and a trowel make planting seeds easier. A trowel is a tool with a pointed scoop for digging holes and loosening soil.

Planting a Crop

The box on this page gives planting instructions for four kinds of crops. Read about each crop. Then decide which crop you'd like to grow.

Tomatoes were brought to Europe from South America in the 1500s, where they were grown in gardens for their pretty fruits. But before the 1800s, many people didn't eat tomatoes. Back then, people thought tomatoes were poisonous.

Corn is actually the grain crop maize, which was first grown in North America over 6000 years ago. After European settlers arrived, they took maize to Europe, where it is an important grain crop too.

Carrots are roots of a plant that was first grown in Europe around the Mediterranean Sea. Carrots were among the first plants brought to North America by explorers.

Strawberries that grow in gardens are probably a hybrid of many kinds of strawberry plants that once grew wild in North and South America.

Seeds of fruits and vegetables are many sizes, shapes, and colors. Which of these seeds can you identify?

Pick a Crop

Tomatoes Start seeds in pots 6 weeks before you'll plant. Set seedlings 46 centimeters apart and slightly deeper than they were growing in the pots.

Corn Plant corn seeds in holes 3 centimeters deep and 5 centimeters apart. Put 4 to 6 seeds in each hole. Thin plants for final spacing of 30 centimeters.

Carrots Plant seeds about 0.6 centimeters deep and in rows that have the soil loosened to the depth of about 20 centimeters.

Strawberries Plant seedlings in holes about 20 centimeters deep. Spread out the roots. Space the plants at least 60 centimeters apart.

Where you plant your crop is an important decision. To choose the best plot for your crop, think about three things. How much sunlight does my crop need? How much space does my crop need? What kind of soil does my crop need?

You should also keep in mind that some plants are hardy—they can grow in places where they might not get all the things they need. Other plants are delicate and won't grow well in places where they don't get enough sunlight or space to grow in.

Before you plant, you should do one more thing—test your soil. A soil test tells how acid or alkaline (al′kə līn) your soil is. Some plants grow best in acid soil while others grow best in alkaline soil. Numbers called pH numbers measure how acid or alkaline the soil is. A pH number below 7 is acid. A pH number above 7 is alkaline. Most plants grow best in soil that is slightly acid—pH 6.5. And few plants can grow at all in soil that is more acid than pH 4 or more alkaline than pH 8.

The dark color of this garden soil is a sign that it is rich in humus.

Pick a Plot

This spot is a sunny, open field, about 12 meters by 8 meters. It's the best plot on your land.

This spot is a patch of sandy soil that is in the shade for part of the day. The sunniest section is about 3 meters by 4 meters.

This plot is a small, gently sloping section, about 3 meters by 1 meter. It gets full sun for most of the day.

This corner section of land gets full sun and has rich soil. Too bad it's only 2 meters by 2 meters.

Caring For a Crop

You've planted your crop. Now you're ready to take care of it! Part of this job is making sure your soil has enough nutrients for growing plants. How? You need to add fertilizer to the soil. Remember that fertilizers contain nutrients that plants need.

Deciding what fertilizer to use isn't easy. Fertilizers have different ingredients and some fertilizers cost more than others do. You can buy chemical fertilizers that are special mixtures for certain types of plants. The packages have labels that list three numbers. The numbers tell the percentages of nitrogen, phosphorus, and potassium they contain. A 5-10-5 label would have 5% nitrogen, 10% phosphorus, and 5% potassium.

You must also decide when to fertilize. Some plants grow better if you add fertilizer before you plant. Other plants need to be fertilized while they are growing. When you add fertilizer, remember this rule: the more nitrogen, the less you should use.

Lettuce needs nitrogen-rich soil to produce dark green leaves.

Pick a Fertilizer

This fertilizer is a 5-10-10 mixture of nutrients. Use it on a crop that needs twice as much phosphorus as nitrogen.

This fertilizer is a 10-20-20 mixture. Use it *before* you plant this delicate crop. Once planted, this crop should not be fertilized.

This 5-10-5 fertilizer is used on a crop that needs lots of nitrogen. Fertilize once a month. Follow the directions to make sure you add enough fertilizer.

This fertilizer is a 10-10-10 mixture. Use it on a crop that needs high amounts of soil nutrients. Fertilize after the crop sprouts and again in a few weeks.

Taking care of your crop also involves protecting it from insect pests. But before you decide how to get rid of your pests, you better make sure they really are "pests"! The praying mantis on this page may not look like something you'd want hanging around your garden. But as you learned, a praying mantis eats many kinds of insect pests. And ladybugs have an enormous appetite for harmful aphids. So keep in mind that you can use other insects to control your garden pests.

You can also use other plants! Some plants make chemicals with odors or tastes that keep insect pests away. Growing these plants next to your crop can help control pests.

Chances are you won't have to use chemical pesticides. A few insects can't eat up all the plants! But pesticides may be needed for large numbers of pests. Then, two rules should be remembered. Follow the label instructions carefully. And it's better to use too little than too much.

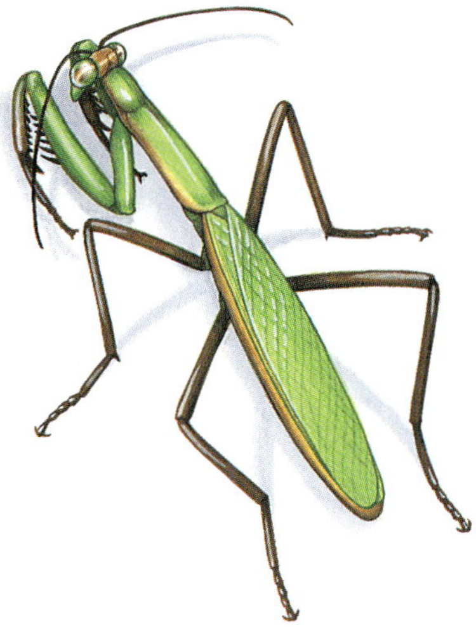

A praying mantis is a predator—a hunter that captures and eats its victim.

Pick a Pest Control Method

Pull out any plants that look sickly to prevent diseases from spreading. Cover the soil with ground-up corn cobs to stop weeds from growing.

Plant the herbs rosemary and sage nearby to help keep pests away. You don't need to use chemical pesticides on this hardy crop.

This crop attracts many pests, so you might need to use a pesticide. Plant onions or potatoes with this crop to help control pests.

Plant asparagus around this crop to keep pests away. Get some ladybugs, which eat harmful aphids. Pick off the hornworm caterpillars.

Harvesting a Crop

If all your choices had the same color box, then your garden grew quite well! Now read about what each crop needs to grow. Once you know more about these crops, play the game again. See if you make better choices this time.

Corn is the crop that needs the most space. It also needs lots of sunlight and nutrient-rich soil. Corn plants must be close together so that wind can pollinate the corn flowers. If wind can't blow pollen from one plant to another, you'll get no ears!

Like corn, tomatoes need a sunny place to grow. But they don't need as much space. Tomatoes grow best in soil that has high amounts of nitrogen and phosphorus.

Carrots are hardy—they grow without special care. Carrots can grow in partial sun and they aren't harmed by many pests. Sandy soil and low amounts of fertilizer are best for carrots. Too much nitrogen can cause carrots to become forked.

Strawberry plants need lots of sunlight, but not a lot of space. In fact, you can even grow strawberry plants in barrels or large flower boxes. Strawberry plants aren't hardy—they often get diseases. Fertilizers should be used *before* planting to prevent damage to the fruit.

Many gardeners make compost in a wire bin, or composter, like the one shown here.

What garden wastes do you see in this compost?

Once fall comes, you may think your job is done. But fall is the time to get your garden ready for next year. A good way to do this is by making **compost**—a kind of natural fertilizer. As you can see, compost is made of the same things humus is made of—dead plant materials. So, like humus, compost is rich in nutrients that plants need.

How do you make compost? At the end of the growing season, pile the dead plants in a big heap. You can also add grass clippings, weeds, leaves, and kitchen scraps, such as apple peelings. Shovel a layer of soil onto the compost heap. Keep adding layers of plant materials and soil whenever you have any garden wastes. As the plant materials decay, compost forms.

When spring arrives, the nutrient-rich compost will be ready to shovel into your garden soil. By adding compost to the soil, you are helping to complete the nutrient cycle you learned about!

Checkpoint

1. How does a soil test help gardeners?
2. Why do gardeners add fertilizer to soil?
3. How does compost improve garden soil?
4. **Take Action!** Look in a garden book or on seed packages for the planting and care instructions for four vegetables. Use the information to design a new garden game.

INVESTIGATE

Struggling Seeds

Will pH affect a seed's growth? Let's investigate.

What to Do

A. Label three cups 1, 2, and 3.

B. Add ten radish seeds to each cup. Then add water to cup 1; vinegar to cup 2; and a seltzer tablet dissolved in water to cup 3.

C. Let all the seeds soak overnight.

D. Remove the seeds from each cup and wrap them in a moist paper towel. Slide each towel into a labeled plastic bag. Seal the bag.

E. Look at the seeds every day for four days. Check for seed growth. The seed is growing when you can see small roots or a shoot. Record the number of seeds that show growth each day.

Record Your Results

Day	Number Growing		
	Cup 1	Cup 2	Cup 3
1			
2			
3			
4			

What Did You Find Out?

1. *Do radish seeds grow in vinegar, which is acid? Do they grow in the solution of the tablet dissolved in water, which is alkaline?*

2. *Why were some of the seeds soaked in water?*

Activity

Passing the Test

Whether you are growing tomatoes or petunias, you need to know if your soil is right for your plants. Try this activity to test your soil's acidity.

Picture A

Picture B

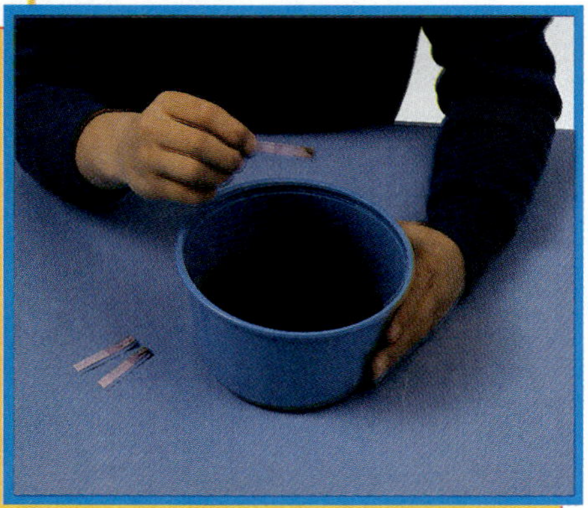

Picture C

Gather These Materials

- soil sample
- clean container
- distilled water
- spoon
- blue litmus paper
- plastic cup

Follow This Procedure

1. Make a chart like the one on the next page. Record your observations in the chart.
2. Collect a soil sample from your yard or school yard. Put the sample in a clean container.
3. Use the spoon to mix the soil sample with enough distilled water for the soil to become muddy, about as thick as applesauce. (Picture A) You will be using litmus paper to test the soil sample. This paper changes color when it is dipped in acid.
4. Insert 3 pieces of blue litmus paper halfway into the muddy soil. The top half of each paper should not get wet. Wait 10 seconds. Then, remove one piece of the litmus paper. Rinse it off with distilled water. (Picture B)
5. Note the color of the litmus paper you removed. If the litmus paper is pink or red, your soil is acid. If the paper is still blue, wait another 5 minutes, and then remove the second piece of litmus paper.

Predict: *Will the litmus paper change color when it is dipped in soil?*

6 If the second piece of litmus paper is pink, this shows that your soil is moderately acid. Record your results. If this second piece is blue, wait another 10 minutes. Then remove the last piece of litmus paper.

7 If the last piece is pink, your soil is slightly acid. If it is still blue, your soil is not acid at all. (Picture C) Record your results.

Record Your Results

	Color of litmus paper
First piece of litmus paper	
Second piece of litmus paper	
Third piece of litmus paper	
Result: Soil sample is ________	

State Your Conclusions

1. What were the results of your soil test?
2. How do you think gardeners could use this type of information when they get ready to grow their plants?

Let's Experiment

What do you think would happen to the acidity of the soil if you add lime to it? Use what you know about scientific methods to find out.

LESSON

3.2 The World's Garden

What's it like to be a farmer?

Gardening is like farming without a lot of worries. Gardeners don't have to make money on their crop. They don't have to pay taxes or borrow money from the bank. Gardening is fun; farming is business. But gardeners and farmers make the same kinds of decisions. So learning how to take care of a garden can help you understand what it's like to take care of a farm.

Farming Today

Suddenly you're in a farmer's boots and blue jeans. You're riding a tractor in an enormous field. Yikes! Your small garden is now a huge farm! What are you going to do now? How much fertilizer will your soil need? What pesticides should you use?

You don't have time or money to test every part of your land for soil pH. So, you'll do what farmers have done for thousands of years. You'll estimate, or use what you know to make a good guess.

What problems does this new fertilizer truck help farmers solve?

You're also going to estimate how much fertilizer your soil needs. And you'll follow this rule—more is better than less. By adding more fertilizer than you think you need, you make sure that even your worst soil has enough nutrients.

Then you remember reading about a new kind of truck that can help you take the guesswork out of soil testing and fertilizing. You make a phone call to the company. Someone from the company brings a truck to your farm and takes you for a ride.

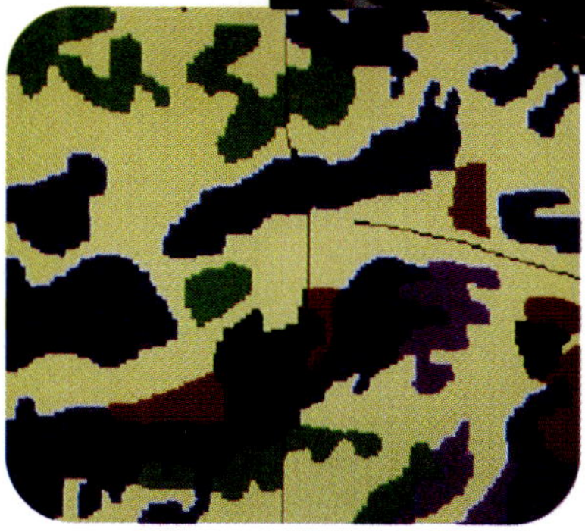

Computerized soil maps, like this one, are analyzed by a computer in the cab of the truck. The different colors represent different kinds of soil.

Wow! This truck is amazing. It has a computer on board. Data about your soil is put into the computer. The data come from soil tests, pictures of your land taken by satellites, and soil maps made by the government. The computer hums as it analyzes the data. Then it chooses the exact amounts of nutrients that are best for your soil. The truck mixes the nutrients together to make a special fertilizer—just for *your* land! A few seconds later, the fertilizer comes pouring out.

In no time at all, this super new machine fertilizes all your fields and your ride is over. You can't believe what it can do! It makes decisions for you. It can help you save money on fertilizer. You'd never use too much fertilizer again. And that would be good news for the environment. You remember how fertilizers polluted a nearby stream. Then you close your eyes and imagine your fields at harvest time. Because your soil has just the right kinds of nutrients, your crops should be larger than ever! Finally, you shake your head and wonder, "Farming sure is changing. What will it be like in the future?"

Into The Field

How does fertilizer affect the growth and health of plants?

Observe several planted areas. Describe the appearance of the plants. Find out whether each planted area is fertilized.

Bypassing the Farm

Someday food may be produced without growing crops.

Imagine growing food without planting crops! A new method may be used to do that.

Tissue culture is a method of growing living tissues outside an organism. To make a tissue culture, scientists remove tissue from an organism and place the tissue in a growing medium. A medium is a liquid that contains growth hormones, water, and nutrients.

Scientists are using tissue cultures to grow cells of crop plants. For example, one scientist made a tissue culture of the juice sacs inside an orange. The juice sacs multiplied on their own and produced orange juice—without any orange trees! In the future, tissue-culture factories, like the ones shown here, may produce foods.

1 Planting trees
Fast-growing trees supply the sugars for growing mediums.

5 Distributing food
The foods are taken to market. Tissue-culture factories can be located anywhere and can operate all year.

2 Harvesting trees
Every year some trees are cut down and shredded into chips.

3 Digesting wood The wood chips are broken down into simple sugars.

4 Growing food
In each factory, the sugars are used in growing mediums for different cells. Corn cells produce corn oil. Wheat cells produce flour. Tomato cells produce tomato juice.

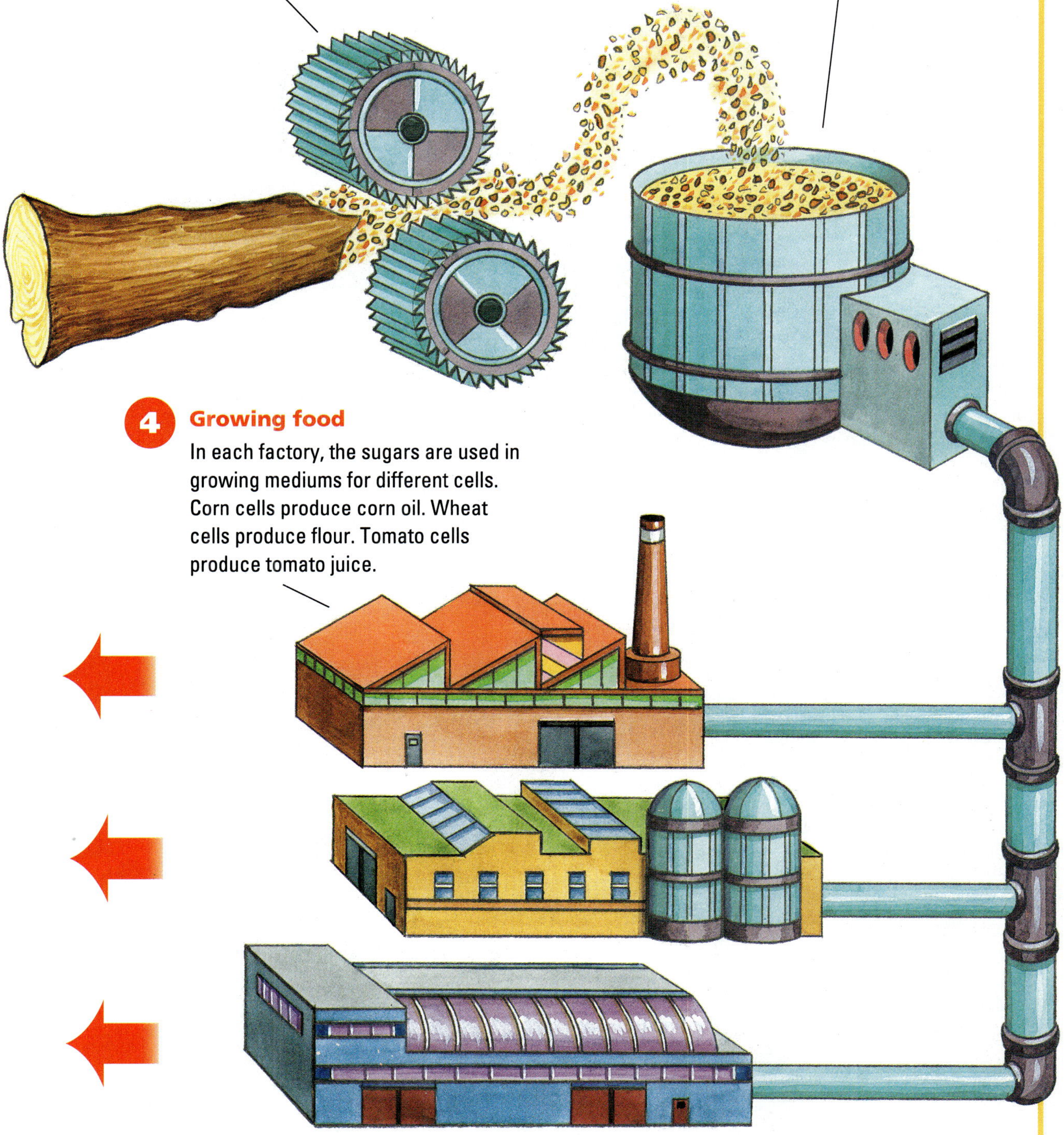

Farming in Space

The year is 2060. You just came from your job in the crop chamber. You and 10,000 other people live and work in a giant wheel spinning in space. As you begin a letter to your grandparents back on Earth, you remember visiting their farm when you were a kid. Things sure were different then!

"Dear Granddad and Grandmother," you write, "You should see how fast plants grow in space! Out here, crops grow all year long. And they get 24 hours of sunlight every day. No wonder we space farmers produce lots of food in our space station! Besides, our crops aren't harmed by any insects. Everything that comes on board our space station is checked to make sure no pests are brought inside. So, we can farm without pesticides.

"And the work is fun! Most of us space farmers are also doing experiments in genetic engineering. I'm developing a new kind of 'super corn.' It grows in hot and cold temperatures. Pretty soon we'll ship some super corn back to Earth. You'll be able to grow corn in winter and in summer."

You close your letter, but add a P.S. "Sometimes I look out at our beautiful planet Earth, and it reminds me of why I'm here. We space farmers don't want to do your job. We can't produce all the food for people on Earth. We just want to find new and better ways for Earth farmers to farm. We're trying to help you grow more food in ways that will protect Earth's soil and water!"

Checkpoint

1. What problems happen when farmers guess how much fertilizer to use?
2. How might tissue culture be used to grow foods in the future?
3. List some benefits of space farming.
4. **Take Action!** Write an advertisement describing a new kind of farm machine.

◄ A space station might have different levels of fields, where corn, soybeans, and other crops would grow.

Activity

Lights On

How well will a plant grow if it is kept in the light for 24 hours a day? Try this activity and see.

Picture A

Picture B

Picture C

Gather These Materials

- 2 young potted plants, same type and size
- masking tape
- marker
- metric ruler
- water
- grow-light or fluorescent lamp
- 2 large identical cardboard boxes

Follow This Procedure

1. Make a chart like the one on the next page. Record your observations in the chart.
2. Use masking tape to label one of your plants A and the other plant B. Measure the height of your 2 plants.
3. Starting in the morning, place both plants in a box. Set up the lamp so that it shines on the plants. (Picture A)
4. Allow the light to shine on the plants for 8 hours.
5. After 8 hours, remove plant A from the box, and put it into an identical box nearby. Cover the box. No light should be shining on plant A when it is in the second box. (Picture B)
6. Allow the light to shine on plant B overnight, giving it a full 24 hours of light.
7. After 24 hours, measure the height of both plants and record the measurements in the chart.

8 Put plant A back into the lighted box. Again, allow the light to shine on plant A for 8 hours and plant B for 24 hours.

***Predict:* Will the number of hours of light affect the growth rate of the plants?**

9 Repeat the procedure every day for 1 to 2 weeks. Water your plants, keeping the soil of each plant equally moist. Measure the height of each plant at the end of each day. (Picture C) Record this information in your chart.

State Your Conclusions

1. Describe any differences you saw in the growth of the 2 plants.
2. Where could you set up a farm where the plants could get light 24 hours a day?
3. If you had a farm here on Earth, what would be the advantages of giving plants 24 hours of light each day? What would be the disadvantages?

Let's Experiment

Now that you have seen how the amount of light can affect plant growth, do you think the type of light could make a difference? Use what you know about scientific methods to find out.

Record Your Results

	Height - Plant A	Height - Plant B
Start		
Day 1		
Day 2		
Day 3		
Day 4		
Day 5		
Day 6		
Day 7		

Chapter Review

Reviewing Words and Concepts

Write the letter of the word or phrase that best completes each sentence.

1. Corn is a _____.
2. A carrot is a _____.
3. Soil with a pH number below 7 is _____.
4. A 5-10-5 fertilizer has 10% _____.
5. Most plants grow best in soil that has a _____ of 6.5.
6. Decayed plant materials make up _____, which is a natural fertilizer.
7. Modern fertilizing trucks use a _____ to determine the right amount and kind of fertilizer for a particular soil.
8. A _____ is living tissue that is grown outside of an organism.
9. When using fertilizers, the rule is: the more _____, the less fertilizer you should use.
10. In space, plants could get 24 hours of _____ each day.

a. pH
b. computer
c. nitrogen
d. root
e. compost
f. sunlight
g. acidic
h. tissue culture
i. grain
j. phosphorus

Connecting Ideas

1. Copy the concept map. Use the terms at the right to complete the map about soil testing.

acid soil **alkaline soil** **pH number**

2. Write a sentence or two that summarizes the ideas shown in the concept map.

Interpreting What You Learned

1. What planting information do you need to know when you plant seeds?
2. What does a soil test tell you?
3. When testing some soil, you find it has a pH number of 3. Is this land suitable for farming? Explain.
4. Identify three elements found in chemical fertilizers.
5. Explain how certain plants can be used to control pests.
6. Would a space farm need pesticides? Explain why or why not.

Performance Assessment

How can acid and alkaline soils be changed?

Materials
- cover goggles
- blue litmus paper
- red litmus paper
- distilled water
- vinegar
- baking soda
- dropper
- 2 plastic bowls
- 4 spoons
- topsoil

Collecting Data

1. Put on your cover goggles. Place three spoonfuls of soil in a clean plastic bowl. Mix a small amount of distilled water with the soil and stir it to make thick mud.
2. To determine whether the soil is acid or alkaline, insert a piece of blue litmus paper and a piece of red litmus paper into the soil. Blue litmus paper will turn red if the soil is acid. Red litmus paper will turn blue if the soil is alkaline. Record whether the soil is acid or alkaline.
3. Add ten drops of vinegar to the soil. Stir the mixture again. Using litmus paper, determine whether the soil is acid or alkaline.
4. Use a clean spoon to add about one-half of a spoonful of baking soda to the soil. Stir the mixture again. Using litmus paper, determine whether the soil is acid or alkaline.
5. Place three spoonfuls of soil in another clean plastic bowl. Mix a small amount of distilled water with the soil and stir it to make thick mud. Use what you know to make alkaline soil. Test with litmus paper to make sure the soil is alkaline. Record your results.
6. Use what you know to turn the alkaline soil to acid soil. Test with litmus paper to make sure the soil is acid.

Analyzing Data

How did you change the alkaline soil to acid soil? How would you change acid soil to alkaline soil?

The Green Teams

City gardens and parks provide open space where people can relax and feel less crowded. The plants and trees help to freshen the city's air. Green leaves and colorful flowers brighten the dull surroundings.

In cities across the United States, people are teaming up to create more green spaces. Students just like you are important members of the teams. Students help improve their communities' environments in many ways. They clear bricks and debris from abandoned lots. They haul dirt and plant flowers. They build playgrounds. They pick up litter and water the grass.

Families from the neighborhood can grow their own flowers and vegetables.

There's a community park in Chicago that's so lovely people call it *El Paraíso,* which means "the paradise" in Spanish. El Paraíso was once an empty lot filled with garbage and rubble. The community stepped in. They didn't want the land to go to waste. So they got permission from the city to turn the site into a park that the whole neighborhood could enjoy. Students and other people in the community worked together. Local businesses donated plants and building materials. Volunteers cleared and fenced the land and planned the layout of the park. They set aside one part for people to sit in and enjoy the surroundings. Another part of the park was set aside as a community garden, where families from the neighborhood can grow their own flowers and vegetables.

In New York City, about 600 older students take part in the Community Volunteer Corps (CVC). These students work on many projects, including ones that help the environment. One CVC project called Lots for Tots has changed several vacant lots into safe playgrounds for small children. CVC volunteers also work in Operation Greenthumb, creating and taking care of public parks. While these students are creating green spaces, they are also training for future jobs.

One CVC project was to reopen a children's garden. First the workers cleared out mounds of rocks, weeds, and trash. Then they built four plant beds and began a garden. In the spring of 1991, the garden produced its first harvest in 30 years! Now students from a local school help keep the garden growing.

Is there a place in your community that could use a spot of green? Maybe a "green team" in your area could use your help. Or perhaps you could start a green team of your own!

On Your Own

To learn what other kids are doing to beautify their communities, write to Renew America. This organization keeps a record of environmental projects going on around the country.

Sweet Potatoes in Space

Dr. Audrey Trotman

Occupation: Agricultural microbiologist

Words to remember: "Science is fun. It's something anyone can do."

Have you ever tried a sweet potato cookie? Have you ever eaten bread made from sweet potato flour? Researchers at Tuskegee University in Alabama are trying to come up with a variety of uses for sweet potatoes and their leaves. Other researchers are trying to find ways to recycle the parts of the plant that cannot be eaten. Still others are studying the best ways to grow sweet potatoes in a hydroponic medium, or nutrient solution. Dr. Audrey Trotman is one of these researchers. She is an agricultural microbiologist—a scientist who studies microorganisms that affect crops.

Why is there so much interest in sweet potatoes?

"Sweet potatoes are just one of the food crops that NASA would like to raise in space. Others include foods such as rice, wheat, white potatoes, soybeans, peanuts, and lettuce. At Tuskegee we're finding out as much as possible about sweet potatoes."

Why do they need to grow plants?

"Some flights, such as the year-long planned flight to Mars, will be too long to carry prepared foods. The answer is to grow the food where it's needed. But in space there will be limits."

What kinds of limits?

"First of all, soil will not be used. We're working with a hydroponic medium that would supply all the nutrients the sweet potato plant needs. One of the main needs of any plant is a source of nitrogen. Some types of bacteria, called nitrogen-fixing bacteria, form compounds that contain nitrogen. I'm working to find how many of these bacteria to add to the hydroponic medium. We're also studying the best way to grow the bacteria in space."

Biosphere 2: Farming Under Glass

Biosphere 2 is a large glass building that is sealed off from the rest of the world. Inside are several miniature environments that resemble places on the earth. Water and wastes are recycled.

1 The rainforest produces oxygen.

2 The ocean supplies water.

3 The wetlands include a beach, a saltwater marsh, and a freshwater marsh.

4 The desert is separated from other areas by dense, thorny shrubs. The savanna is a grassland area.

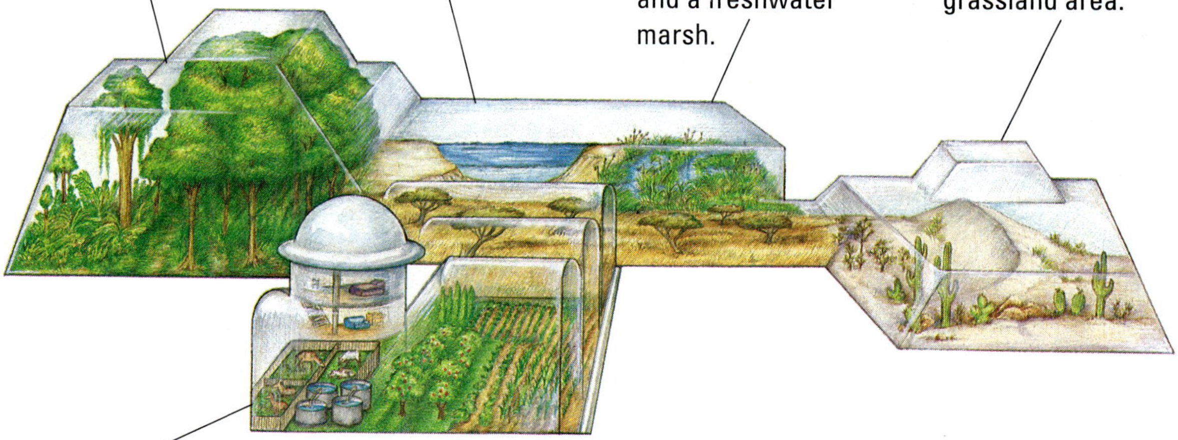

5 The agricultural section contains living quarters for eight people as well as a farm. Biosphere 2 residents raise fish in large tanks. Nitrogen-rich water from the fish tanks is used to fertilize crops, which are grown by hydroponics. Ladybugs and wasps are used instead of chemicals to control crop pests.

Find Out On Your Own

Check current science magazines for reports on the progress of the Biosphere 2 experiment. Find magazine articles in the library by using the *Readers' Guide to Periodical Literature*.

Module Performance Assessment

Botanical Garden

Using what you learned in this module, design activities for a botanical garden. A botanical garden is like a museum of living plants. Use one of the following activities to design one of the stations where students will stop to learn different things about plants. You may work by yourself or with others in a group.

History

People did not always cultivate plants for food. For thousands of years, people simply gathered plants to eat that grew wild. Research the history of agriculture. Prepare a time line showing some highlights in the development of agriculture from earliest times to the present.

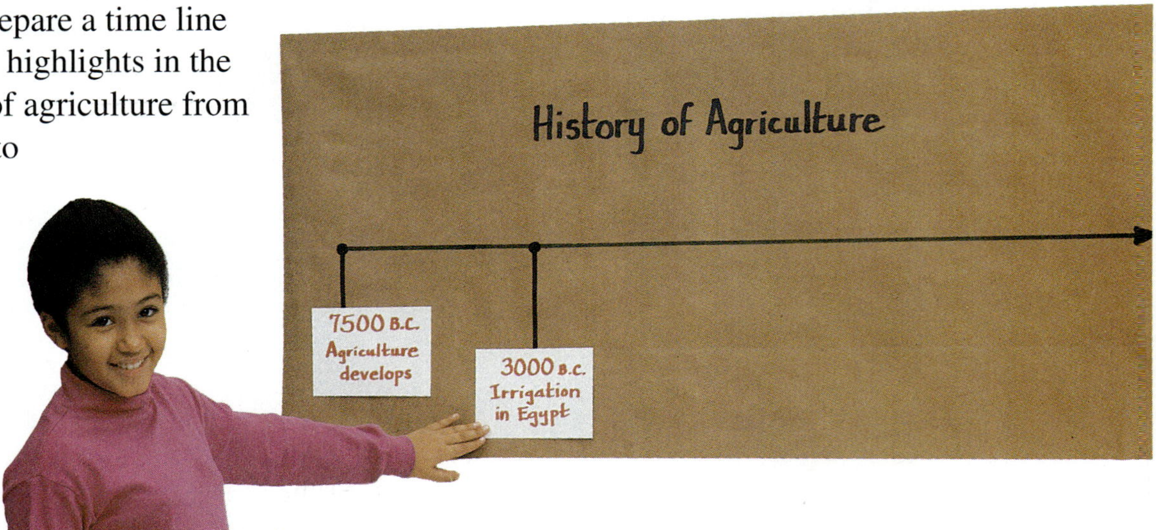

Biology

Use seedlings of two different kinds of plants, a fluorescent lamp, water, soil with a pH of 5, and a 10-20-20 fertilizer to show how each kind of plant has its own special needs. Make a drawing to show what this station would look like.

Art

Design a poster that illustrates the steps that might produce a genetically engineered corn plant that can change nitrogen gas into nitrogen compounds. Use short, easy-to-read paragraphs to explain each step.

Social Studies

Your local legislators are considering a law to ban the use of chemical pesticides in your community. Design a display that shows arguments favoring and opposing such legislation. Consider the views of farmers, health professionals, chemical pesticide manufacturers and their employees, people who fish in local waters, ordinary citizens, and elected officials.

Drama

Prepare a skit that shows how plants communicate according to the "plant talk" hypothesis. Some students should portray insect pests. Other students should portray plants being attacked by the insect pests. Perform the skit for your classmates. Have them guess which steps of the plant talk hypothesis are being portrayed.

Module Review

Reviewing Words and Concepts

Write the letter of the word or phrase that best completes each sentence.

1. A ____ is a chemical messenger that regulates a process.
2. A ____ helps living things of the same species communicate with each other.
3. A ____ can be used to replace a chemical pesticide.
4. A fertilizer adds ____ to poor soil.
5. A hydroponic farmer uses ____ instead of soil to grow plants.
6. Pollination occurs in plants when ____ moves from a stamen to a pistil.
7. A gene is an inherited bit of information that controls a ____.
8. Genetic engineering can be used to change the traits of an organism by changing its ____.
9. Compost is a natural ____.
10. A tissue ____ is made of cells that are grown outside of an organism.

a. pollen
b. pheromone
c. fertilizer
d. trait
e. culture
f. nutrients
g. water
h. genes
i. biological control
j. hormone

Interpreting What You Learned

1. Explain the effect that harvesting crops has on the cycling of nutrients.
2. Describe how hormones cause the changes that occur when an apple seed develops into a tree.
3. Methods of dealing with crop pests have changed over time. List several reasons why these methods have changed.
4. How did the choices you made in the garden game model those you would make when you grow house plants?

Applying What You Learned

1. What would pH readings of 4 and 8 tell you about two different samples of soil?
2. How do the traits of a hybrid compare to those of the parent plants?
3. Suppose a disease greatly reduced the population of *Rhizobium* bacteria. What effect might this have on plant populations?
4. How would the development of genetically engineered plants that could make nitrogen compounds benefit the environment?

Using Metric

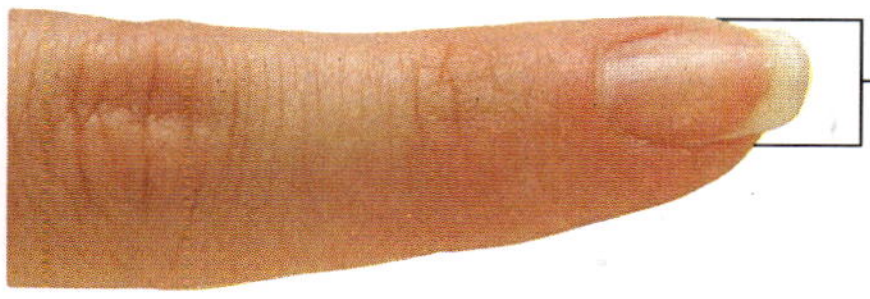

About 1 centimeter

About 1 millimeter

About 1 meter

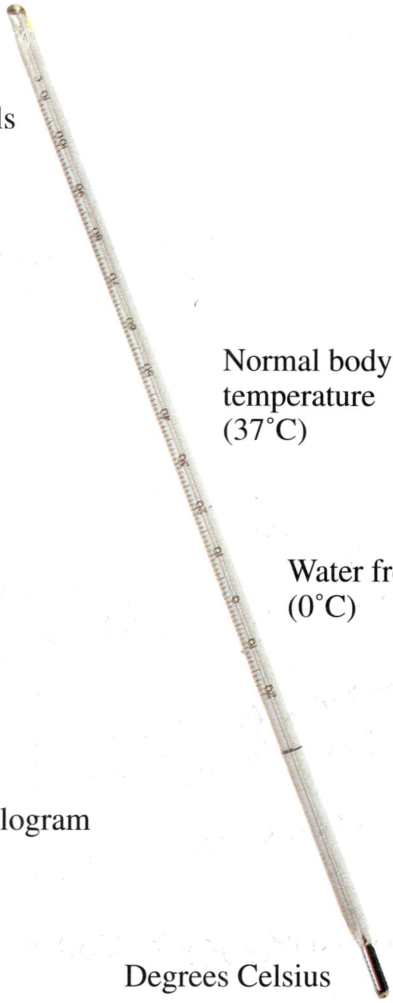

Water boils (100°C)

Normal body temperature (37°C)

Water freezes (0°C)

Degrees Celsius

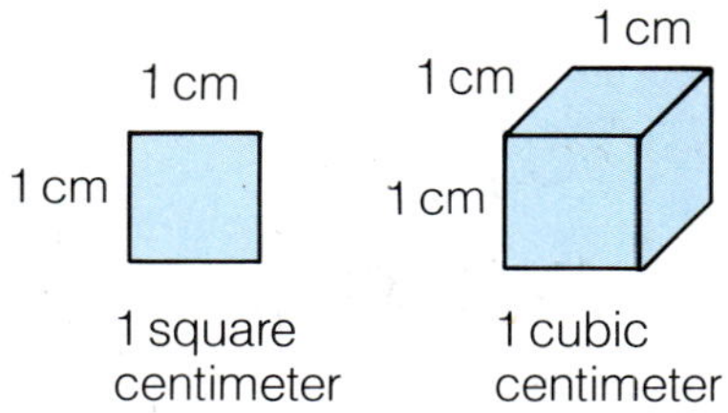

About 1 kilogram

11 football fields end to end is about 1 kilometer

1 liter of milk

Using Scientific Methods

Scientists ask many questions. The answers cannot always be found in books. In fact, no one may know the answers. Then scientists use scientific methods to find answers. Scientific methods include steps like those described on the next page. Sometimes the order of the steps changes. The experiments in this section give you a chance to use scientific methods to explore the world around you.

Identify Problem The problem is usually in the form of a question such as, "How does wind speed affect erosion?"

Make Observations Recorded observations become data and might include the size, color, or shape of something.

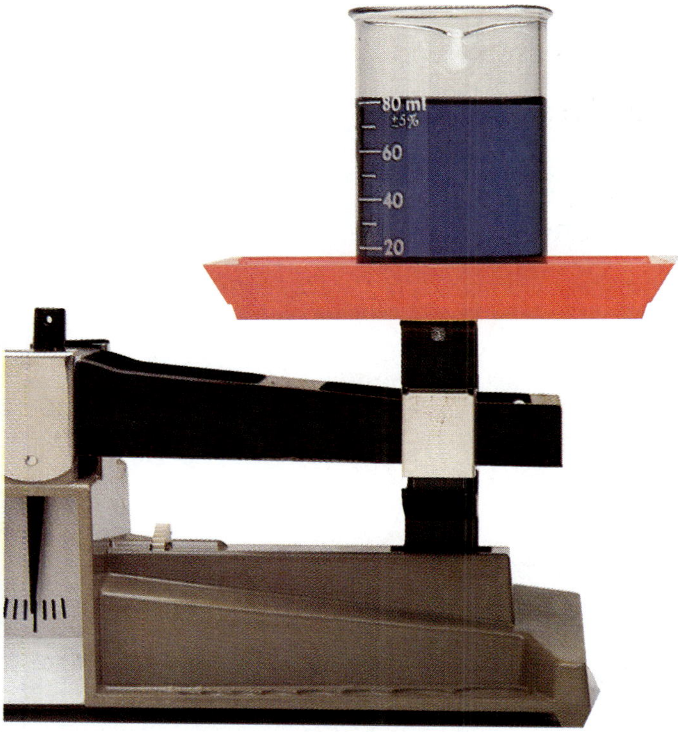

State Hypothesis A hypothesis is a likely explanation of the problem. It may turn out to be incorrect; it must be tested.

Test Hypothesis If possible, experiments are done to test the hypothesis. Experiments should be repeated to double check the results.

Collect Data The information you gather from the experiment is your data.

Study Data The data collected during an experiment is better understood if it is organized into charts and graphs. Then you can easily see what it all means.

Make Conclusions The conclusion relates to the hypothesis. You might conclude your hypothesis is correct, or that it is incorrect.

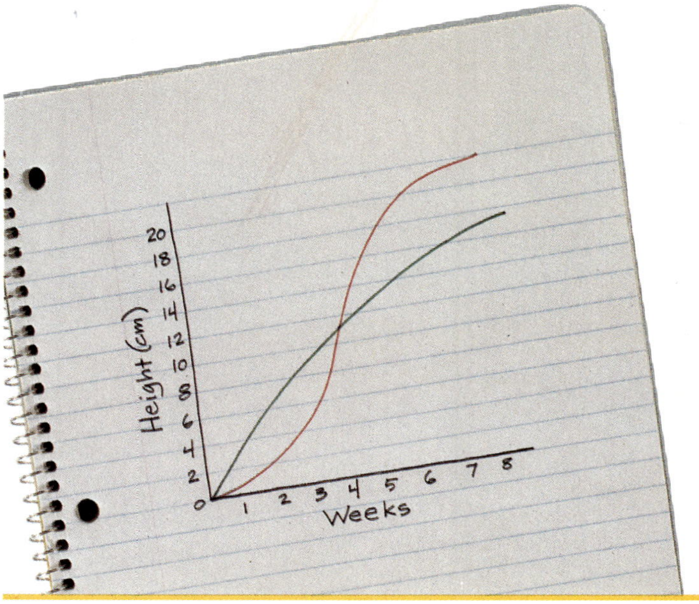

Safety in Science

Scientists know they must work safely when doing experiments. You need to be careful when doing experiments too. Here are some safety tips to remember.

Safety Tips

- Read each experiment carefully.
- Wear cover goggles when needed.
- Clean up spills right away.
- Never taste or smell substances unless directed to do so by your teacher.
- Handle sharp items carefully.
- Tape sharp edges of materials.
- Handle thermometers carefully.
- Put materials away when you finish an experiment.
- Wash your hands after each experiment.

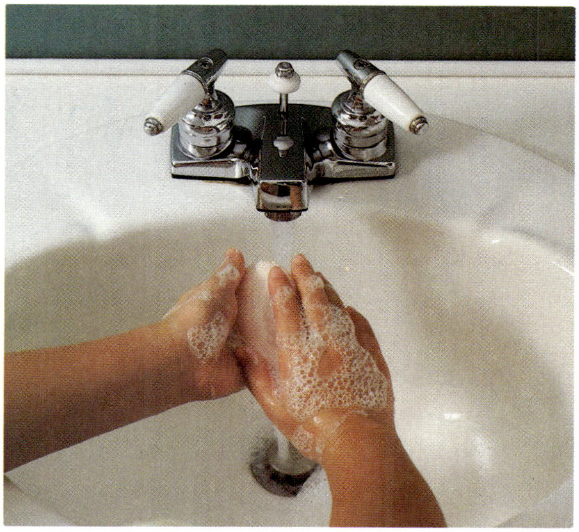

Experiment Skills

Making Conclusions

MODULE A

Experimenting with Dyes for Cell Parts

Maria observed an onion cell through a microscope during her science class. She learned that cells had a nucleus and a cell wall. Maria wondered if cell parts would show up more clearly when stained with dye.

With her teacher's permission, Maria set up an experiment using black ink and green food coloring as dyes. She put a thin piece of onion skin on each of three microscope slides. She added a drop of black ink to one slide, a drop of food coloring to another, and a drop of water to the third. Then she put a coverslip on each slide.

Maria observed the slides under a microscope. She saw the nuclei of the cells on the slide with black ink very clearly. Nuclei did not show up as well on the slide with green food coloring. Nuclei of cells from the slide with water looked the same color as the rest of the cell. Maria studied her results to decide what they meant. She concluded that cell parts show up more clearly when stained with black ink.

Thinking About the Experiment

Maria used her observations to make a conclusion. Every experiment must be set up so that a conclusion can be drawn after the observations are made or data are collected. The conclusion must be supported by the data.

1. What was Maria's conclusion?

2. Which observations led Maria to her conclusion?

Anything in the experiment that could be changed is a variable. A control is the part of the experiment that does not have the variable being tested. The slide with only water was a control.

3. In Maria's experiment, what was the variable that changed from one setup to the other?

4. How did the control help Maria reach her conclusion?

Try It!

Try Maria's experiment and see if you come to the same conclusion.

Problem

What substance makes cell parts more visible under a microscope?

Hypothesis

Dye will make cell parts show up more clearly.

Materials

black ink
green food coloring
water
medicine dropper
microscope
3 microscope slides with coverslips
3 thin pieces of onion skin
tweezers

Procedure

1. Use the tweezers to pull 3 thin pieces of onion skin from a fresh onion bulb.
2. Place a piece of onion skin on each of the microscope slides.
3. With the medicine dropper, place a drop of water on one slide, a drop of green food coloring on another, and a drop of black ink on the third.

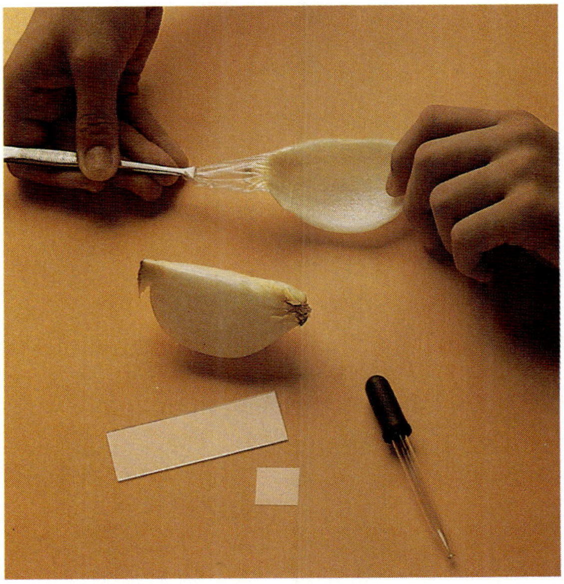

4. Cover each slide with a coverslip.
5. Examine the 3 slides under a microscope.
6. Record your observations on a table like the one below.

Data and Observations

Substance added	Visible cell parts

Conclusion

Write your conclusion based on your data and observations.

Practice

Making Conclusions

Scientists use a purple dye called Gram stain to classify bacteria into two major groups. Suppose you added Gram stain to two kinds of bacteria. One kind turned purple. The other kind did not turn purple.

1. What conclusion can you draw about how cells of different bacteria absorb this stain?
2. What conclusion can you draw about how the bacteria can be classified?

Experiment Skills

Collecting Data

MODULE B

Experimenting with Balloon Rockets

Jim and his father were looking at a rocket engine in a space museum. Jim noticed the opening where gases leave the engine. He wondered if the size of the opening had anything to do with the engine's power.

Jim remembered a time when he was washing his family's car. There was a spray nozzle on the end of the hose. As he changed the size of the opening of the nozzle, Jim had to hold the hose tightly to keep it from twisting in his hand because the force of the water became stronger. If he turned the nozzle so that its opening changed size again, the water did not have as much force.

Jim thought that a rocket engine with a bigger opening might push the rocket farther than an engine with a small opening. He decided to do an experiment to test his hypothesis.

Thinking About the Experiment

Jim wanted to collect data that would help him test his hypothesis. In the procedure on the next page, Jim had to change only one variable.

1. What was the variable that Jim tested?
2. What did Jim's data describe?

In collecting data, Jim recorded that the balloon moved about 2.5 meters in the first test. In the second test, it moved about 3.5 meters. In the third test, it moved about 5 meters.

3. Which nozzle size caused the balloon to move the greatest distance?

Based on the data he collected, Jim made a conclusion about his hypothesis.

4. What was Jim's hypothesis?
5. Based on the data in questions 3 and 4, what do you think Jim concluded about his hypothesis?

Try It!

Try Jim's experiment and see if you come to the same conclusion.

Problem

How does the size of the balloon opening affect how far a balloon rocket travels?

Hypothesis

A balloon rocket with a large opening will travel farther than a balloon rocket with a small opening.

Materials

cover goggles	cardboard, at least 6 cm x 6 cm
8 meters of nylon fishing line	pencil
balloon	masking tape
paper clip	meter stick
scissors	straw

Procedure

1. Cut a 5 cm circle from cardboard.
2. Using the pencil point, make a very small hole in the center of the cardboard.
3. Thread the open end of a balloon through the hole and blow up the balloon. Measure the length of the balloon.
4. Use a paper clip to close the opening of the balloon.
5. Tape a straw to the top of the balloon so that one end of the straw is pointed in the same direction as the opening.

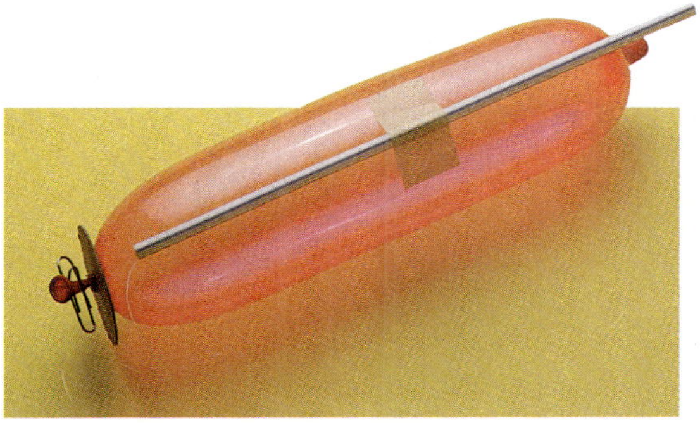

6. Thread the fishing line through the straw. Have a partner hold one end. You hold the other. Stretch it out.
7. Remove the paper clip from the opening. Measure and record how far the balloon moves along the string.
8. Make the hole in the cardboard circle larger. Inflate the balloon to the same size as in step 3. Repeat steps 4-7.
9. Remove the cardboard circle. Inflate the balloon again to the same size. The opening will be at its biggest this time. Repeat steps 4-7.

Data and Observations

Opening size	Distance balloon moves
Small	
Bigger	
Biggest	

Conclusion

Write your conclusion based on your data and observations.

Practice

Collecting Data

Suppose you wanted to collect data about how balloon size affects the distance traveled.

1. What data would you collect about the size of the balloon?
2. What might your hypothesis be?
3. How would you change this experiment so that you could collect data to test your hypothesis?

Experiment Skills

Testing a Hypothesis

MODULE B

Experimenting with Solids

Sally watched her parents move into their new house. Her father could not fit a bookcase through a door. He took the bookcase apart. It then fit easily through the door.

Sally thought of a hypothesis to explain what she saw. Her hypothesis was that the parts of any object take up less space than the object itself does. She decided she would do an experiment to test her hypothesis.

Sally wanted to use a solid in the experiment because the bookcase is a solid. At school, she had used water to measure volume. Volume tells how much space something takes up. Sally used clay for the solid and water to measure volume.

First she measured the volume of a ball of clay. Then she measured the volume of the clay pieces.

Thinking About the Experiment

Sally poured a volume of 250 milliliters of water into the cup. The difference in the water level when a solid is added to the water is the volume of the solid. The water level went up to 300 milliliters with the clay ball and up to 300 milliliters with the clay pieces.

1. What was the volume of water Sally used?

2. What did the volume rise to with the clay ball? the clay pieces?

3. What was the volume of the clay ball? the clay pieces? (Hint: The volume of the clay is the amount the water level changed.)

Sally's experiment showed that her hypothesis was not true for clay.

4. Explain why Sally's data did not support her hypothesis.

5. Write a new hypothesis that supports Sally's data.

Try It!

Try Sally's experiment and see if you come to the same conclusion.

Problem

How does the space taken up by a solid compare to the space taken up by its pieces?

Hypothesis

Recall that Sally's hypothesis was incorrect. Write your own hypothesis for this experiment.

Materials

500 mL measuring cup
metric ruler
water
clay

Procedure

1. Fill the measuring cup exactly to the 250 mL mark.
2. Use a ruler to make a ball of modeling clay that is about 5 cm in diameter.
3. Put the clay into the water. Record the water level.
4. Take the clay out and break it into small pieces.
5. Make certain that the water level is still at 250 mL.
6. Put all of the pieces into the water. Record the water level.

Data and Observations

	Volume in mL
Water	
Clay ball and water	
Clay pieces and water	

Conclusion

Write your conclusion based on your data and observations.

Practice

Testing a Hypothesis

Sally plans to try the same test on a variety of other solids. She wants to start with a block of wood and its pieces.

1. How can she change the experiment so it will work for wood?
2. Suggest some other solids she might use in the experiment.

Experiment Skills

Using Models

MODULE C

Experimenting with Bones

Sheila visited a skeleton display at a science museum. She noticed that the upper body and lower body seemed to meet at a bone shaped something like a triangle. A sign told Sheila that this bone was called the sacrum. Sheila thought that the sacrum must be very strong because it could hold up the other parts of the body. She wondered if other shapes, a square for example, could support as much weight as a triangle does. When she got home Sheila made a model to find out.

She decided to use flexible plastic straws for her model because she could bend them into shapes. She made a triangle and a square from the straws. To see which was stronger, she taped each shape to the end of a ruler. Then she placed books on the ruler one at a time. Sheila found that the triangle could hold up more books than the square before losing its shape.

Thinking About the Experiment

1. What problem did Sheila want to solve?
2. Write a hypothesis for the problem.
3. Sheila used a model to test her hypothesis instead of using real bones. Why?
4. Why did Sheila use flexible straws for her model?
5. How was her model like the real bone?
6. How was the model different?
7. When is it useful to use a model?
8. What conclusion did Sheila make using her model?

Sheila wanted to find out more about bones. She wanted to know what minerals are important to grow strong bones.

9. Could she use her model to answer this question? Why or why not?

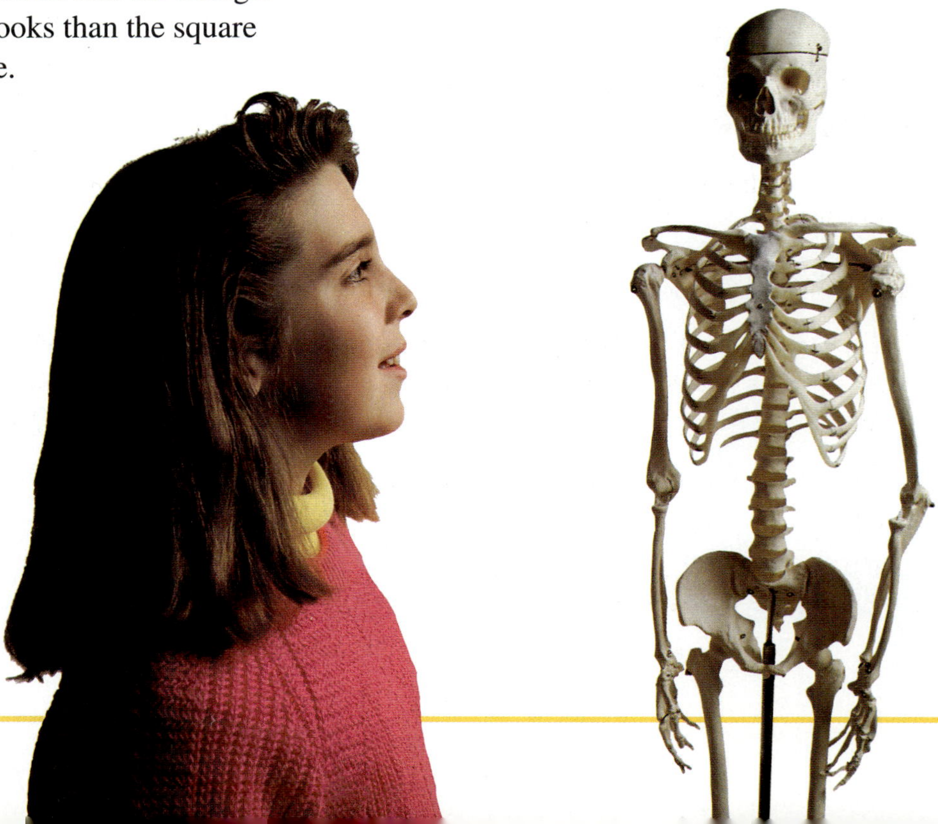

Try It!

Try Sheila's experiment and see if you come to the same conclusion.

Problem

Does a bone shaped like a triangle support weight better than a bone shaped like a square?

Hypothesis

Write your own hypothesis for this experiment.

Materials

- several books of about the same size and weight
- masking tape
- 7 flexible plastic straws
- metric ruler
- scissors

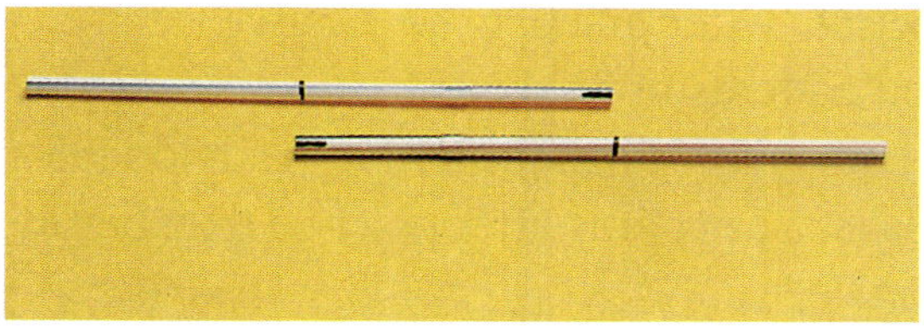

Procedure

1. Trim the long ends of the straws so that the flexible part is in the center. Stretch out the flexible part.
2. Cut 2 slits about 1 cm long at either end of each straw.
3. Use 3 of the straws to make a triangle by fitting the ends together.
4. Use the other 4 straws to make a square.
5. Tape one side of the triangle and one side of the square to the table so that they stand upright.
6. Hold the ruler so that one end is on the table and the other end is supported by the top of the triangle. Fasten the ruler to the triangle with tape.

7. Place a book on the ruler. Keep adding books until the triangle begins to come apart.
8. Repeat steps 6 and 7 with the square. Compare the difference in the strength of the two shapes.

Data and Observations

Shape	Number of books supported
Triangle	
Square	

Conclusion

Write your conclusion based on your data and observations.

Practice

Using Models

Suppose you wanted to show how the bones work inside a joint, such as the shoulder or the knee. Look at a picture of a skeleton to get an idea.

1. What would the model have to do to show how the joint works?
2. Probably several different materials could be used to make a model of a joint. What materials can you think of?
3. How would your model be similar to the real body part?
4. How would it be different?

Experiment Skills

Collecting Data

MODULE C

Experimenting with Practice

Tina was taking piano lessons. She knew that the more she practiced a song, the faster and better she could play it. She wondered if practicing could help a person do any activity faster.

Tina asked Andy to help her find out. Tina traced Andy's right hand on a piece of paper. On the drawing, Tina numbered the fingers 1 through 5 with Andy's thumb being 1. Then Tina asked Andy to tap his fingers in the order 43512. Tina timed how long Andy took to tap this order five times. She wrote the time on a chart. Next, Tina told Andy to practice tapping 43512 for two minutes. Andy practiced. Tina timed him again.

Next, Tina traced Andy's left hand and numbered the fingers. She had Andy tap in the order 25134 five times. Tina wrote down the time. Then she had Andy practice with his left hand for two minutes. Finally, Tina timed Andy again and wrote down the time.

Thinking About the Experiment

Tina was collecting data each time she wrote down how long Andy took to do a finger pattern five times.

1. Why was it important for Tina to write down her data after each step?
2. How is using a data chart like the one on the next page helpful?

Tina's data showed that Andy did five right-hand patterns in 15 seconds before practice. After practice, he did five right-hand patterns in 5 seconds.

3. How many times faster did Andy do the right-hand patterns after practice than before practice?

Tina's data showed that, with practice, Andy cut his left-hand time from 10 seconds to 5 seconds.

4. For which hand did Andy show the most improvement?
5. Did all the data Tina collected support her hypothesis? Explain.
6. Based on his data, what conclusion do you think Tina made?

Try It!

Try Tina's experiment and see if you come to the same conclusion.

Problem

Does practice decrease the time needed to do an activity?

Hypothesis

Practice decreases the time needed to do an activity.

Materials

paper
pen
watch or clock with a second hand

Procedure

1. Trace your partner's right hand on a piece of paper.
2. On the drawing, number the fingers 1 to 5 starting with the thumb.

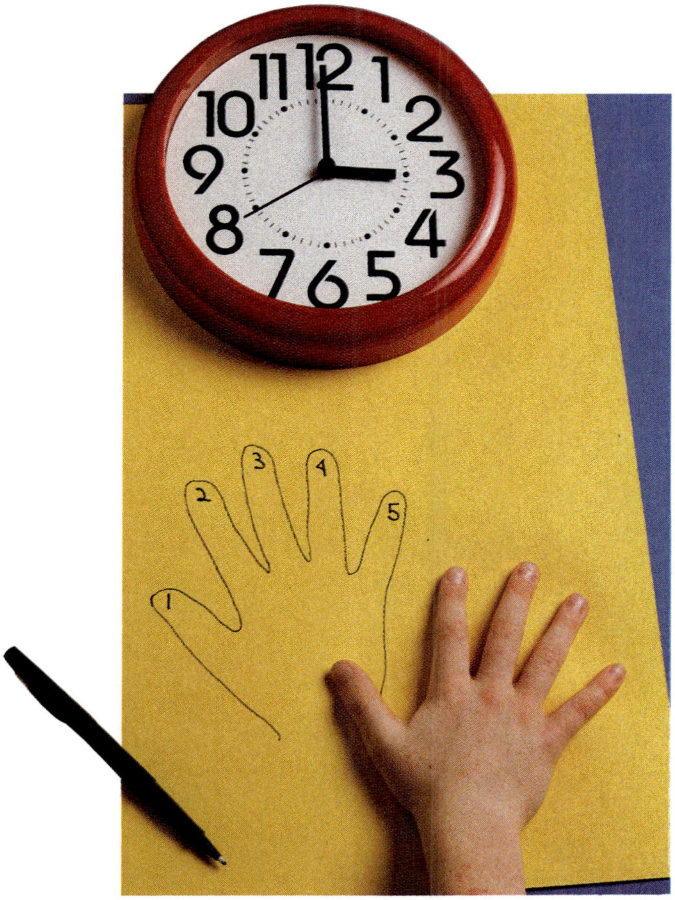

3. Ask your partner to tap his or her fingers in the order 43512 while you time the activity.
4. Record the time your partner needed to tap 43512 five times in a chart like the one below.
5. Have your partner practice tapping the pattern for 2 minutes.
6. Repeat steps 3 and 4.
7. Repeat steps 1-6 using the left hand and having your partner tap in the order 25134.

Data and Observations

	Right hand	Left hand
Time before practice		
Time after practice		

Conclusion

Write your conclusion based on your data and observations.

Practice

Collecting Data

1. Tina wondered how Andy would do if he stopped practicing for a few hours and then tried to do the patterns again. What might Tina's hypothesis be?
2. How would you change the experiment so that you could collect data on how stopping practice affects learning?

Experiment Skills

Studying Data

MODULE D

Experimenting with Water and Sand

Mike lived near a large lake. Sometimes in summer he spent a day at the beach. He noticed that in the morning, the sand on the beach was cool. In the afternoon, the sand was so hot it burned his feet. Mike wondered why the water in the lake did not heat up during the day as the sand did. He thought of this hypothesis to answer his question: *Under a heat source, sand heats up faster than water.*

Mike decided to try an experiment to test his hypothesis. He carefully collected and studied his data. Then he made conclusions about whether sand heats faster than water.

Thinking About the Experiment

The purpose of an experiment is to collect data about the problem being solved. To help study the data, it can be arranged in a table.

1. What kind of data would Mike want to collect in his experiment?

2. Why would Mike want to collect his data at regular times?

3. What should Mike do with his data to help study it?

4. How would studying the data help Mike test his hypothesis?

Mike could use the information in his data table to make a graph. A graph is very useful for showing changes that occur during an experiment. A graph has two lines that cross, or intersect. Each of the two lines in a graph is used to show one kind of measurement data.

5. What information can Mike show on each of the two lines in a graph?

Try It!

Try Mike's experiment and see if you come to the same conclusion.

Problem

Why does lake water not heat up during a warm day like sand does?

Hypothesis

Sand heats up faster than water.

Materials

- 2 aluminum pie pans
- water at room temperature
- sand
- 2 thermometers
- meter stick
- clock or watch with second hand
- 2 200-watt lamps with reflectors and stands

Procedure

1 Fill a pie pan with water. Fill the other pan to the same level with sand.

2 Place the pans about a meter apart on a flat surface. Let the pans stand for about 2 hours to be sure they are at the same temperature.

3 Place a thermometer at the same depth in the middle of each pan. Record the temperatures in the pans in a chart like the one shown.

4 Place the lamps about 40 cm above the pans. Direct the light straight down onto the pans.

5 Measure and record the temperature of each pan every 5 minutes for the next 30 minutes.

Data and Observations

Time (minutes)	Temperature	
	Water	Sand
0		
5		
10		
15		
20		
25		
30		

Conclusion

Write your conclusion based on your data and observations.

Practice

Studying Data

Suppose you wanted to find out whether sand or water cools down faster.

1. What might be your hypothesis?
2. How would you set up an experiment to test your hypothesis?
3. What data would you want to collect and study?
4. How would you organize your data?

Experiment Skills

Making Observations

MODULE E

Experimenting with Magnetic Fields

Juanita was studying magnets in science. She was very interested in the shapes of magnetic fields. She remembered a picture in her science text that showed how iron filings lined up around a magnet. Juanita wondered if magnets of different shapes had magnetic fields of different shapes. She set up an experiment to find out.

Juanita placed some iron powder in a jar that contained cooking oil. She shook the jar and held it over a magnet. The iron powder lined up to show a three-dimensional pattern of the magnetic field. Juanita made careful observations of the magnetic field. She tried several different magnets. Juanita was able to compare the magnetic fields of magnets of different shapes by making detailed drawings.

Thinking About the Experiment

One important part of making observations is to be able to notice differences. Juanita observed that iron powder seemed to build up on the ends of one of the magnets she tested.

1. What does this observation indicate?

When making good observations it is important to notice not only changes, but also where changes do not occur. Juanita noticed that the iron powder near the top of the oil was not affected by the magnet.

2. Is this an important observation for Juanita to make? Explain.

When making observations it is important to make an accurate record of what you observe. Juanita decided to record her observations by making a drawing of what she saw.

3. How else might Juanita have recorded her observations?

Try It!

Try Juanita's experiment and see if you come to the same conclusion.

Problem

How does the shape of a magnet affect the shape of its magnetic field?

Hypothesis

Magnets of different shapes have magnetic fields of different shapes.

Materials

small plastic jar with lid	iron powder
vegetable oil	magnets of different shapes
teaspoon	

Procedure

1. Fill the plastic jar to about 2.5 cm from the top with clear vegetable oil.
2. Add one teaspoon of powdered iron to the oil. Put on the lid and make sure that it is tightly closed. Shake the jar to mix the iron and oil.
3. Hold the jar steady over each type of magnet you wish to test. Notice what happens to the particles of iron in the oil.
4. Record your observations by making drawings. Draw the pattern of what you see around the magnets you are using.

Data and Observations

Magnetic field pattern

Conclusion

Write your conclusion based on your data and observations.

Practice

Making Observations

The experiment allows you to observe the magnetic field on the part of the magnet where the plastic bottle is held.

1. How would you change the experiment to enable you to observe the magnetic field all around the magnet?
2. Why might it be important to observe the magnetic field all around the magnet?

Experiment Skills

Collecting Data

MODULE E

Experimenting with an Electromagnet

After studying electromagnets in science, Nathan and Mike decided to do an experiment with electromagnets. They each used a book that told them how to build an electromagnet.

The boys tested their electromagnets by trying to pick up small iron and steel objects. Nathan's electromagnet seemed stronger than Mike's. Mike noticed that Nathan was using two batteries, while he was using only one. Mike wondered if the number of batteries made a difference in an electromagnet's strength. The boys decided to do an experiment to find out.

Thinking About the Experiment

In order to collect data that would help them reach a conclusion, Nathan and Mike had to choose one variable to test.

1. What variable did Nathan and Mike test?

2. Why would Nathan and Mike want to use only paper clips instead of several different iron and steel items?

Each time the boys added a battery to the electromagnet, they collected new data. The one-battery magnet picked up three clips. The two-battery magnet picked up seven clips. The three-battery magnet picked up ten clips.

3. Which electromagnet was weakest?

4. Which electromagnet was strongest?

Nathan and Mike concluded that an electromagnet is made stronger by adding more batteries to it.

5. Does Nathan and Mike's data support their conclusion? Explain your answer.

6. Suppose Nathan and Mike had added a fourth battery to their electromagnet and tested it. Would the electromagnet have picked up more or fewer clips than it did with three batteries? Explain your answer.

Try It!

Try Nathan and Mike's experiment and see if you come to the same conclusion.

Problem

What affects the strength of an electromagnet?

Hypothesis

Adding batteries to an electromagnet makes it stronger.

Materials

3 flashlight batteries
insulated copper wire
sandpaper
large nail
small paper clips
masking tape

Procedure

1. Use sandpaper to remove 1 cm of insulation from the wire's ends.
2. Wind the wire around the nail at least 30 times. Do not let the wire coils overlap.

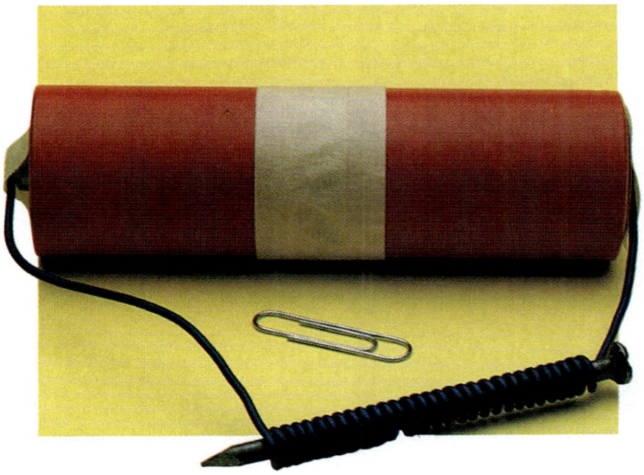

3. Tape 1 end of the wire to 1 end of a battery. Tape the other end of the wire to the other end of the battery.
4. Try to pick up paper clips with the nail. Count and record the number of clips that the magnet picks up.
5. Take the ends of the wire off the battery.
6. Connect 2 batteries by placing the positive (+) end of 1 battery to the negative (-) end of another. Tape the batteries together tightly.
7. Tape the ends of the wire to the free ends of the batteries.
8. Repeat steps 4 and 5.
9. Take the wires off the batteries and add a third battery to the series. Make sure you only connect a positive end with a negative end.
10. Repeat steps 4 and 5.

Data and Observations

Batteries	Paper clips
1	
2	
3	

Conclusion

Write your conclusion based on your data and observations.

Practice

Collecting Data

1. Suppose you wanted to test how the size of the nail affects the strength of an electromagnet. How would you change the experiment above?
2. Instead of collecting data about the number of clips picked up by different numbers of batteries, what data would you collect?

Experiment Skills

Making Conclusions

MODULE E

Experimenting with a Wet Cell

Alicia had learned that the current in a wet cell battery is caused by electrons flowing through a liquid containing an acid. She knew that some foods have juices that contain acids. She wondered if some foods contain strong enough acids to make a wet cell battery. She decided to do an experiment to find out.

Alicia's teacher told her that a compass needle will move when an electric current flows near it. Alicia wrapped wire around a compass. She taped one end of the wire to a copper strip and the other end to a zinc strip. She pushed the two metal strips into a hot dog bun. She observed the compass needle to see if it moved. It did not.

Alicia repeated this procedure three times, using a lemon, a head of lettuce, and an orange.

Thinking About the Experiment

Alicia used the data she collected to make conclusions. Alicia recorded that the needle moved when she used the orange and the lemon. It did not move when she used the hot dog bun or the lettuce.

1. With what foods did current flow?

2. What proof did Alicia have that current flowed with the foods listed in question 1?

Alicia concluded that some citrus fruits contain acids strong enough to make a wet cell battery.

3. Was Alicia's conclusion correct? Explain your answer.

Lemons and oranges are citrus fruits.

4. If Alicia had concluded that all citrus fruits work in a wet cell battery, would you agree with her conclusion? Explain your answer.

Try It!

Try Alicia's experiment and see if you come to the same conclusion.

Problem

Is the acid in some foods strong enough to be used to make a wet cell battery?

Hypothesis

Some foods contain strong enough acids to make a wet cell battery.

Materials

compass
insulated copper wire
cover goggles
masking tape
copper strip, 2 cm wide, 10 cm long
zinc strip 2 cm wide, 10 cm long
lemon
hot dog bun
head of lettuce
orange
sandpaper
pencil
metric ruler

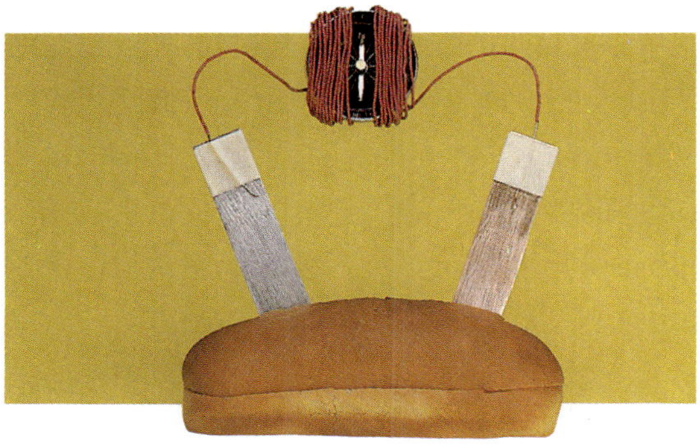

Procedure

1. Put on your cover goggles. Rub the metal strips with sandpaper until the strips are shiny.
2. Use the sandpaper to remove 3 cm of insulation from each end of the wire.
3. Wrap the wire around the compass 25 times. Make sure the wire loops all go in the same direction.
4. Tape one end of the wire to the copper strip. Tape the other end of the wire to the zinc strip.
5. Turn the compass so that the needle is in line with the wire coils.
6. Make a small pencil mark 3 cm from the bottom of each metal strip.
7. Make two small pencil marks 4 cm apart on the bun. Push the ends of the metal strips 3 cm into the bun.
8. Observe the compass needle and record your observation.
9. Press and roll the lemon on a table or desk top. Repeat steps 7-8 using the lemon.
10. Repeat steps 7-8 using the lettuce.
11. Press and roll the orange. Repeat steps 7-8 using the orange.

Data and Observations

Food used	Needle moves	Current flows
Hot dog bun		
Lemon		
Lettuce		
Orange		

Conclusion

Write your conclusion based on your data and observations.

Practice

Making Conclusions

Alicia hypothesized that other metals can be used to make a lemon wet cell.

1. Describe an experiment that Alicia could do to test her hypothesis.
2. What kind of data would Alicia need to collect to make a conclusion about her hypothesis?

Experiment Skills

Setting Up a Control

MODULE F

Experimenting with Plant Responses

Kristi wanted to have a plant for her room. She made a cutting from a large plant and placed it in a jar of water. After a few days, she saw roots growing straight downward from the plant into the water. Kristi wondered if plant roots always grow downward, with gravity. Kristi wondered what would happen if she could turn a plant upside down. Would the roots bend and start to grow in the opposite direction? She did an experiment to find out.

Kristi used corn seeds that she soaked in water. She put the seeds on paper towels placed against the inside of two jars containing some water. In a few days, the young roots grew straight downward. Kristi turned one jar upside down so the roots pointed upward. A few days later she noticed the roots bent all the way around so they were starting to grow downward. She compared them to the roots of the seedlings in the upright jar. The seeds in the upright jar had grown straight down without bending.

Thinking About the Experiment

Kristi did something to Jar A that she did not do to Jar B. The part that is different is the variable being tested.

1. What is the variable being tested?

Kristi's experiment also had a control. A control is part of the experiment that does not have the variable.

2. Which group of seedlings were the control?
3. How did having a control help Kristi see the results of the experiment?
4. Suppose Kristi did not use control plants. Could she be sure that turning the plants caused the roots to change direction? Why or why not?
5. Suppose Kristi forgot to add water to the control plants. How would this affect her results?
6. Why should the control in the experiment have all conditions the same *except* for the tested variable?

Try It!

Try Kristi's experiment and see if you come to the same conclusion.

Problem

How do plant roots respond to gravity?

Hypothesis

Plant roots always grow downward with gravity.

Materials

2 pieces of cardboard
8 corn seeds
dish
graduated cylinder
2 jars with lids
marker
paper towel
water

Procedure

1. Place the corn seeds in a dish of water. Soak them overnight.
2. Mark the two jar lids *A* and *B*.
3. Fold a paper towel so that it fits flat around the inside of each jar. Wedge a piece of cardboard up against it so that the paper towel stays up against the side of the jar.

4. In each jar, put 4 soaked corn seeds between the paper towel and the side of the jar. Place the seeds so the pointed ends are downward. The paper towel should keep the seeds in place.
5. Add 50 mL water to the bottom of each jar and cover.
6. Keep both jars in a dark place.
7. After 3 days, observe the direction of the root growth of the seedlings in each jar.
8. Place Jar *A* on its side. Make sure the lid is screwed on tightly. Leave Jar *B* upright.
9. Observe both jars after another 3 days.
10. Place Jar *A* so it is upside down. Observe again after 3 days.

Data and Observations

Observations of Root Growth

Day	Jar A	Jar B
3		
6		
9		

Conclusion

Write your conclusion based on your data and observations.

Practice

Setting up a Control

Suppose you wanted to find out if plant stems always grow upward. You might use three potted plants. One pot could be placed on its side. Another would hang from a rope so it is upside down.

1. What should you do with the third pot?
2. What is the variable being tested in your experiment?
3. What is the control in your experiment?

Experiment Skills

Collecting Data

MODULE F

Experimenting with Plant Shoots

One day Erin watched her aunt snip buds off the tops of plants she had grown in pots on her windowsill. She told Erin that removing the top bud caused the plants to grow new buds and shoots. These buds and shoots grew from the sides of the main stem. If she did not snip the top buds, she explained, the plants would grow tall but would have fewer side shoots.

Erin set up an experiment to test how removing buds from plants would affect their growth. She used six plants growing in small pots. The plants ranged between twelve centimeters and fifteen centimeters in height, and each had nearly the same number of shoots growing from the sides of the main stem.

Erin snipped the top bud off three of the plants and did nothing to the other three plants. After ten days she counted the number of side shoots on each plant and measured each plant's height.

Thinking About the Experiment

Erin used only one kind of plant in her experiment. Suppose she had compared different kinds of plants.

1. Explain why this would have made the experiment more difficult for Erin to make conclusions from her data.

Erin collected data about six plants—three with the top bud removed and three without the top bud removed.

2. Why did Erin not use just one plant with a top bud and one without?

3. Explain why Erin did not remove the top bud from plants in pots 1 to 3.

4. Why did Erin choose plants that were about the same height and had about the same number of stems at the start of the experiment?

5. Why was it necessary to put all of the plants on the same sunny windowsill?

Try It!

Try Erin's experiment and see if you come to the same conclusion.

Problem

How does removing the bud from the top of some plants affect their growth?

Hypothesis

Some plants grow new shoots from the sides of the main stem after a top bud is removed.

Materials

- 6 potted plants, each 12 cm to 15 cm tall
- metric ruler
- masking tape
- marker

Procedure

1. Place a piece of masking tape on the side of each pot. Use the tape to label the pots, *1, 2,* and *3.*
2. Label the other pots *4, 5,* and *6.*
3. Using your fingertips pinch off the top bud on the three plants in pots *4-6.*

4. Measure the heights of all of the plants and record these in the data table.
5. Count the number of side shoots on each plant. Record the numbers in the data table.
6. Place the plants on a sunny windowsill and let them stay there for 8 days. Do not forget to water them regularly.
7. After 8 days, measure the height of each plant and record the measurements in the data table.
8. Count the number of side shoots on each plant and record these in the data table.

Data and Observations

	Plants					
Height	1	2	3	4	5	6
Day 1						
Day 8						
Shoots						
Day 1						
Day 8						

Conclusion

Write your conclusion based on your data and observations.

Practice

Collecting Data

Suppose you wanted to see how removing the end buds from all the side stems of a plant affected its growth.

1. What kind of data might you collect?
2. What characteristics might you look for in a plant to make collecting your data easier?
3. Make a data table that you might use to record your observations.

Experiment Skills

Setting Up an Experiment

MODULE F

Experimenting with Plants

Chang learned that plants are adapted to different climates. He thought that water probably passes out of a forest plant faster than it does from a desert plant. He collected a green leaf from a tree in his yard and a small piece of a jade plant. A jade plant has small, thick leaves and is similar to a desert plant.

Chang set up an experiment. He put water in three clear plastic cups. He hung the tree leaf over one cup and the piece of jade plant over another cup so that each was in water. Thus, both could absorb water. The third cup contained only water. He covered each of the three cups with another cup.

Chang put the three pairs of cups in the same place. He observed the top cups for several days.

Thinking About the Experiment

Chang wanted to compare the rate at which water was released to the air from each plant.

1. What is the purpose of the upper cups?
2. What did Chang expect to see in the upper cups?

For his comparison, Chang needed a control. The control is the part of the set-up that is not affected by what happens to the leaves.

3. Which part of the setup was the control in this experiment?
4. Why was the control important in Chang's experiment?
5. Why did Chang leave all three setups in the same location?

Try It!

Try Chang's experiment and see if you come to the same conclusion.

Problem

Does water pass out of a forest plant faster than from a desert plant?

Hypothesis

Forest plants release water more quickly than do desert plants.

Materials

green leaf
6 clear plastic cups
small stem with leaves from a jade plant
nail
3 squares of cardboard large enough to cover the cups
tape
water

Procedure

1. Fill 3 of the plastic cups two-thirds full with water.
2. Use a nail to pierce the center of each cardboard square.

3. Place the plant stems through two of the cardboard squares.
4. Put each square on top of a cup. Make sure the two plant stems are in the water.
5. Use tape to hold the cardboard onto the cups.
6. Place an empty cup over each square. Tape it securely to the cardboard.
7. Place the three setups in the same location .
8. Observe the top cups for several days.

Data and Observations

	Day 1	Day 2
Forest Leaf		
Jade plant		
No plant		

Conclusion

Write your conclusion based on your data and observations.

Practice

Setting Up an Experiment

Explain how you would change the experiment above to find out each of the following.

1. What could you do to measure the amount of water absorbed by each plant?
2. What could you do to find out how temperature affects the amount of water released by a plant?

Glossary

Pronunciation Key

The pronunciation of each word is shown just after the word, in this way: **ab bre vi ate** (ə brē′vē āt). The letters and signs used are pronounced as in the words to the right. The mark ′ is placed after a syllable with primary or heavy accent, as in the example above. The mark ′ after a syllable shows a secondary or lighter accent, as in **ab bre vi a tion** (ə brē′vē ā′shən).

a	hat, cap	j	jam, enjoy	TH	then, smooth		
ā	age, face	k	kind, seek				
ä	father, far	l	land, coal	u	cup, butter		
		m	me, am	ů	full, put		
b	bad, rob	n	no, in	ü	rule, move		
ch	child, much	ng	long, bring				
d	did, red			v	very, save		
		o	hot, rock	w	will, woman		
e	let, best	ō	open, go	y	young, yet		
ē	equal, be	ȯ	order, all	z	zero, breeze		
ėr	term, learn	oi	oil, voice	zh	measure, seizure		
		ou	house, out				
f	fat, if						
g	go, bag	p	paper, cup	ə	represents:		
h	he, how	r	run, try		a in about		
		s	say, yes		e in taken		
i	it, pin	sh	she, rush		i in pencil		
ī	ice, five	t	tell, it		o in lemon		
		th	thin, both		u in circus		

A

acceleration (ak sel′ə rā′shən), the rate of change of velocity.

acid (as′id), substance with a pH lower than 7; a compound that releases hydrogen ions in water.

air resistance (ar ri zis′təns), the opposition to motion caused by air particles.

alga (al′gə), one of a group of freshwater or saltwater producers classified in the protist kingdom.

alkaline (al′kə līn), refers to a base; describes a substance with a pH higher than 7.

animalcule (an′ə mal′kyül), name used by Leeuwenhoek to describe the microscopic organisms he observed.

antenna (an ten′ə), device for receiving and sending radio waves.

atmosphere (at′me sfir), layered mixture of gases surrounding the earth.

atom (at′əm), basic unit of all matter.

atomic number, the number of protons in the nucleus of the atom.

autonomic (ô′tə nom′ik) **nervous system,** controls involuntary processes such as breathing and digestion.

axon (ak′son), part of the nerve cell that carries impulses away from the cell body.

B

bacteria (bak tir′ē ə), group of microscopic nonphotosynthetic one-celled organisms classified in the moneran kingdom.

ball and socket joint, one of the five kinds of joints in the body; allows circular motion.
bile (bīl), a bitter, greenish-yellow liquid that aids digestion; released by the liver.
biological (bī′ə loj′ə kəl) **control,** method that uses natural enemies to control insects.

C

Calorie (kal′ər ē), measure of the amount of energy in foods; the amount of energy needed to raise the temperature of one kilogram of water one degree Celsius.
carbon dioxide (kär′bən dī ok′sīd), a gas in air; each molecule consists of one atom of carbon, two atoms of oxygen.
carnivore (kär′nə vôr), consumer that eats only other consumers.
cartilage (kär′tl ij), tough, elastic, flexible material that covers bones in joints to prevent friction and absorb vibrations.
cell (sel), the basic living unit of an organism.
CFC, abbreviation for chlorofluorocarbon gas; a gas in the atmosphere that can break down ozone molecules.
chemical (kem′ə kəl) **bond,** the strong attraction that holds two atoms together in a substance.
chemical energy, energy stored in the chemical bonds between atoms and molecules.
chemical equation (i kwā′zhən), an arrangement of symbols, formulas, and numbers that are used to describe a chemical reaction.
chemical reaction, a process that produces one or more new substances.
chlorophyll (klôr′ə fil), green substance in producers that absorbs light energy for photosynthesis.
chloroplast (klôr′ə plast), organelle in producers that contains chlorophyll.
chromosome (krō′mə sōm), threadlike strand that is made up of genes; found in the nucleus of a cell.
circuit (sėr′kit), a closed path through which current flows.
circuit breaker, safety device that automatically opens an electric circuit when too much current flows through it.
circulatory (sėr′kyə lə tôr ē) **system,** the network of arteries and veins that carries blood throughout an organism.
classify (klas′ə fī), to sort into groups based on similarities and differences.
cleavage (klē′vij), the way a crystal splits into pieces with flat surfaces.
cochlea (kok′lē ə), spiral shaped structure in the inner ear filled with fluid and lined with tiny, hairlike cells.
complex machine, machine made up of many simple and compound machines that is usually powered by electricity or fuel.
compost (kom′pōst), mixture of decaying leaves, manure, and other material that forms a kind of natural fertilizer.
compound (kom′pound), substance produced when atoms of two or more elements combine.
compound machine, a machine that is a combination of two or more simple machines.
comsat (kom′sat′), a communications satellite.
conductor (kən duk′ter), a material that transmits heat or electricity.
consumer (kən sü′mər), organisms that get their energy from the stored chemical energy in the food they eat.
control (kən trōl′), the part of an experiment that does not have the variable being tested.
cornea (kôr′nē ə), the transparent part of the outer coat of the eyeball, covering the iris and pupil.

crest (krest), highest point on a wave.

crystal (kris′tl), structural unit in most solids that has an orderly, repeating arrangement.

cytoplasm (sī tə plaz′əm), clear, jellylike fluid substance between the nuclear membrane and the cell membrane; contains all the cell's organelles and aids in moving material within the cell.

D

DDT, abbreviation for dichloro-diphenyl-trichloro-ethane, a powerful insecticide.

decomposer (dē′kəm pō′zer), organism that helps to break down and decay dead organisms and the wastes of living organisms.

dendrite (den′drīt), part of a nerve cell where nerve impulses begin and then travel through to the cell body.

depressant (di pres′nt), kind of drug, such as alcohol and tranquilizers, that slows down the nervous system and reduces reaction time.

digestion (də jes′chen), the breaking down of food into forms animals can use.

drag, the disturbance of air or water caused by motion that slows the body down.

E

eardrum, the thin skin that covers the end of the ear canal.

echo (ek′ō), reflected sound waves.

egg (eg), female sex cell.

electric (i lek′trik) **current,** the flow of electric charge.

electrical (i lek′trə kəl) **energy,** kind of energy carried by moving charges.

electromagnetic (i lek′trō mag net′ik) **wave,** energy that travels through space at a speed of 300,000,000 meters per second.

electromagnetic radiation (rā′də ā′shən), energy moving in waves through space.

electromagnetic spectrum (spek′trəm), scale of electromagnetic waves from shortest to longest or longest to shortest.

electron (i lek′tron), negatively charged particle that moves around the nucleus of an atom.

electron cloud model, current model of the atom that states that the exact location of the electrons is unknown.

element (el′ə mənt), any substance made of only one kind of atom; one of 109 basic substances that cannot be broken down by chemical reactions.

energy (en′ər jē), the ability to do work or change matter; usually expressed in joules.

enzyme (en′zīm), substance that controls how quickly chemical reactions occur in the body.

epithelial (ep′ə thē′lē əl) **cell,** kind of cell that forms the skin and linings of organs such as the mouth and intestines.

esophagus (ē sof′ə gəs), the tube that carries food and fluids from the mouth to the stomach.

eukaryote (yü kar′ē ōt), organism made of cells that contain many different structures surrounded by thin coverings called membranes.

F

farsighted (fär′sī′tid), kind of sight where images form beyond the retina and are not seen clearly.

fertilizer (fėr′tl ī′zər), substance, either chemical or natural, spread over or put into the soil to supply missing nutrients.

fixed joint, one of five kinds of joints in the body; does not allow movement.

food chain, simple feeding pattern that shows the transfer of energy from one organism to another.

food web, system of food chains in an ecosystem that shows how energy flows through the ecosystem.

force (fôrs), a push or a pull.

friction (frik′shən), force between surfaces that resists the motion of one surface past another surface.

fruit (früt), thickened ovary that covers and protects the seed or seeds.

fulcrum (ful′krəm), point on which a lever is supported and turns.

fungicide (fun′jə sīd), any chemical that kills fungus.

fuse (fyüz), safety device in which a metal strip melts to open a circuit when too much current flows through it.

G

gastric (gas′ trik) **juice,** strong acid made in the stomach that aids digestion.

gear (gir), toothed wheel that developed from the wheel and the lever.

gene (jēn), a unit of information that controls a trait.

genetic engineering (jə net′ik en′jə nir′ing), a technology to change the traits of an organism by changing its genes.

geode (jē′ōd), rock usually having a cavity lined with crystals or other mineral matter.

geographic (jē′ə graf′ik) **pole,** a fixed location at either end of the earth's axis.

gliding joint, one of five kinds of joints in the body; allows bones to slide past one another.

gravity (grav′ə tē), a force that exists between any two objects.

greenhouse effect, the trapping of infrared radiation from the earth by some gases.

greenhouse gas, gas in the atmosphere that traps heat near the earth's surface instead of allowing it to pass through to space.

H

HDTV, abbreviation for High Definition Television, a technology that allows a larger, clearer picture than other forms of television.

hemoglobin (hē′mə glō′bən), red-colored protein in red blood cells that carries oxygen to cells.

herbicide (hėr′bə sīd), chemical used to kill weeds.

herbivore (hėr′bə vôr), a consumer that eats only producers.

hinge joint, one of five kinds of joints in the body; allows back-and-forth movement.

hormone (hôr′mōn), chemical substance that affects growth and development of organisms.

horsepower, customary unit that measures power.

humus (hyü′məs), part of soil formed from decayed plant-and-animal matter.

hybrid (hī′brid), an offspring of parents with different traits.

hydroponics (hī′drə pon′iks), the science of growing plants in a nutrient-rich liquid instead of soil.

hypothesis (hī poth′ə sis), a likely explanation of a problem.

I

inclined (in klīnd′) **plane**, a simple machine that is a flat surface with one end higher than the other.

inertia (in ėr′shə), the tendency of a moving object to stay in motion or a resting object to stay still.

insulator (in′sə lā′tər), a material that does not conduct heat or electricity well.

integrated pest management, use of a combination of ways to control insects and other pests.

integumentary (in teg′yə men′tər ē) **system,** the system that covers and protects the body.

involuntary (in vol′ən ter′ē) **muscle,** a muscle that works without a person thinking about it.

iris (ī′ris), colored ring-shaped muscles in the eye that change the size of the pupil.

J

joint, place where bones come together.

joule (joul), metric unit used for measuring work and energy.

K

kinetic (ki net′ik) **energy,** energy of motion.

L

larva (lär′və), immature caterpillar stage of many insects.

legume (leg′yüm), certain plants whose roots have nodules that contain nitrogen-fixing bacteria.

lens, a transparent oval body in the eye directly behind the iris, that focuses light rays upon the retina.

lever (lev′ər), a simple machine made of a bar that is held up on a point, or support, called a fulcrum.

ligament (lig′ə mənt), tough strips of tissue that hold bones together at joints.

lightning (līt′ning), giant spark of moving electric charges in the atmosphere.

lubricate (lü′brə kāt), to apply oil, grease, or other substance to reduce friction.

luster (lus′tər), property that tells how a substance reflects light.

M

macrophage (mak′rō fāj), kind of white blood cell capable of absorbing and destroying waste or harmful material.

magnetic (mag net′ik) **pole,** place on a magnet where the magnetic field is strongest.

magnetic field, the space around a magnet in which an object feels a magnetic force.

magnetism (mag′nətiz′əm), the force of attraction that comes from a magnet.

marrow (mar′ō), soft tissue inside some bones that produces red blood cells.

mass (mas), the amount of material that an object has in it.

mechanical (mə kan′ə kəl) **energy,** energy an object has because of its motion and the forces acting on it.

medium (mē′dē əm), liquid that contains growth hormones, water, and nutrients.

metal (met′l), one of a group of elements that are usually solid at room temperature, are shiny, and conduct heat and electricity well.

microorganism (mī′krō ôr′gə niz′əm), organisms too small to be seen with the unaided eye.

mineral (min′ər əl), basic, solid, nonliving matter from the earth.

mitochondrion (mī tə kon′drē ən), organelle that releases energy in the cell.

model (mod′l), a picture or structure that will help to explain how something looks or works.

molecule (mol′ə kyül), the smallest bit of a compound that has the same properties of that compound.

muscular (mus′kyə lər) **system,** the muscles attached to your skeleton.

N

NASA, abbreviation for the National Aeronautics and Space Administration.

nearsighted (nir′sī′tid), kind of sight where images form in front of the retina and cannot be seen clearly.

nerve impulses, signals that relay messages between your brain and your muscles.

nervous system, system that links all parts of the body and carries information back and forth among them.

neutron (nü′tron), particle having no electric charge and found in the nucleus of the atom.

nodule (noj′ül), a small structure on the roots of legumes that contains nitrogen-fixing bacteria.

nonmetal (non met′l), one of a group of elements that cannot be polished, is brittle, and does not conduct heat or electricity.

nuclear membrane (nü′klē ər mem′brān), covering that surrounds the nucleus.

nucleus (nü′klē əs), central part of the atom; control center of the cell.

nutrient (nü′trē ənt), a substance needed for growth and survival.

O

omnivore (äm′ni′vōər), a consumer that eats both producers and consumers.

optic (op′tik) **nerve,** bundle of nerves connecting the retina to the brain.

orbit (ôr′bit), the path of an object around another object.

organ (ôr′gən), groups of tissues that work together to do a job.

organ system, groups of organs that work together to do a job.

organelle (ôr′gə nel′), structures found in the cytoplasm that do the work of the cell.

organism (ôr′gə niz′əm), a living thing made up of one cell or more.

ovule (ō′vyül), contains the female sex cell.

ozone (ō′zōn), kind of oxygen that contains three oxygen atoms per molecule.

P

pancreas (pan′krē əs), a large gland near the stomach that gives off pancreatic juice, a digestive juice which contains various enzymes, into the small intestine.

parallel (par′ə lel) **circuit,** circuit that allows the current to flow along more than one path.

periodic (pir′ē od′ik) **table,** chart that shows the orderly arrangements of the different elements based on their atomic numbers.

pesticide (pes′tə sīd), chemical that kills pests that damage crops.

pH number, number from 0-14 that tells whether a substance is acid or alkaline.

pheromone (fer ə mōn), any chemical substance that is given off by members of a species and causes other members of the species to give particular responses.

photosynthesis (fō′tō sin′thə sis), the process by which producers make sugars from sunlight, water, and carbon dioxide.

pistil (pis′tl), female part of a flower that contains the ovules.

pivot joint, one of five kinds of joints in the body; allows bones to move over and under one another.

plasmid (plaz′mid), ringlike structure of a bacterium that contains genetic material.

polarizing (pō′lə rīz′ing) **lens,** a lens that allows electromagnetic waves vibrating in one direction only to pass through.

pollen (pol′ən), tiny grain that contains the male sex cell.

pollination (pol′ə nā′shən), the movement of pollen from a stamen to a pistil.

potential (pə ten′shəl) **energy,** stored energy an object has because of its position.

power (pou′ər), rate at which work is done.

power surge (sėrj), a sharp, brief increase in the amount of current in a wire.

producer (prə dü′sər), an organism that makes its own sugars using the sun's energy.

prokaryote (prō kar′ē ōt), organism made of cells without a membrane separating organelles from the cytoplasm.

proton (prō′ton), positively charged particle in the nucleus of an atom.

pulley (pu̇l′ē), a simple machine consisting of a rope that passes over a grooved wheel.

pupil (pyü′pəl), opening into the eye through which light passes.

pyrethrum (pī rē′thrəm), chemical poison made of the powdered flower heads of any of various chrysanthemums.

Q

quark (kwôrk), a basic building block of the nucleus of an atom.

R

reflex (rē′fleks), any movement that happens automatically.

repel (ri pel′), to push away.

resistance (ri zis′təns), anything that slows down the flow of electric charge.

respiration (res′pə rā′shən), the process by which cells change sugar and oxygen into carbon dioxide and water, thereby releasing energy.

retina (ret′n ə), a layer of nerve cells in the back of the eye that contains both the rods and cones.

rolling friction, friction that results when a round object rolls over a surface.

S

saliva (sə lī′və), watery fluid in the mouth that moistens food, aids chewing and swallowing, and begins digestion.

satellite (sat′l īt), object that revolves, or moves, around another object.

scale (skāl), the size of a plan, map, drawing, or model compared with what it represents.

scanning electron microscope, tool that forms an image by sending a beam of electrons across surfaces of objects.

screw (skrü), inclined plane in a spiral form.

seed (sēd), the result of an egg and sperm joining together.

series (sir′ēz) **circuit,** circuit in which current can move only through one pathway.

simple machine (mə shēn′), tool with only one or two parts.

skeletal (skel′ə təl) **system,** the framework of bones and other parts; supports and protects the body.

sliding friction, friction that results when two solid surfaces slide across one another.

small intestine (in tes′tən), the organ of the digestive system in which most digestion takes place.

smog (smog), pollution consisting of a combination of smoke and fog in the air.

solar (sō′lər) **cell,** device that changes solar energy directly into electrical energy.

solar energy, energy from the sun.

sound energy from vibrating matter.

species (spē′shēz), a group of organisms that have similar traits and can produce offspring that can also produce offspring.

speed, measures how quickly an object moves over a certain distance.

sperm (spėrm), male sex cell.

spongy bone, layer of bone with an open, spongelike structure that makes bones lighter and easier to move.

sprocket (sprok′it), gear on a bicycle.

stamen (stā′mən), the male part of a flower that produces pollen.

stimulant (stim′yə lənt), kind of drug, such as cocaine and amphetamines, that speed up the nervous system.

stomach (stum′ək), storage sac where most food begins the digestion process.

streamlining (strēm′lin′ing), rounding of the surfaces of objects to reduce air resistance.

sunspot (sun′spot′), dark patch on the sun where a magnetic storm is occurring.

symbol (sim′bəl), one or two letters that stand for an element's name.

synapse (si naps′), gap between the dendrites of one nerve cell and the axon of the next where a nerve impulse passes from one nerve cell to another.

T

tendon (ten′dən), ropelike structure that connects muscles to bones.

theory (thē′ər ē), one or more related hypotheses supported by data that best explains things or events.

tissue (tish′ü), groups of cells of the same kind that work together to do a job.

tissue culture (kul′chər), a method of growing living tissues outside an organism.

tolerance (tol′ər əns), a user's need to take more and more of a drug to feel an effect.

trait (trāt), a particular characteristic or property of an organism.

transmission electron microscope, forms an image by transmitting a beam of electrons through thinly-sliced objects.

transmitter (tran smit′ər), antenna that sends out radio waves.

tread (tred), the grooves and ridges on a bicycle tire.

troposphere (trop′ə sfir), lowest level of the atmosphere from sea level up to about 11 kilometers.

trough (trôf), lowest point of a wave.

V

vacuole (vak′yü ōl), organelle of storage in the cell.

variable (ver′ē ə bəl), anything in an experiment that can be changed.

velocity (və los′ə tē), value giving the speed and direction in which an object is moving.

vibration (vī brā′shən), rapid back-and-forth motion.

villus (vil′əs), tiny fingerlike projection in the small intestine that increases the area for absorption of food.

visible (viz′ə bəl) **light,** light that humans can see.

voluntary (vol′ən ter′ē) **muscle,** muscle that a person can control with thought.

W

watt (wot), metric unit that measures power.

wavelength (wāv′lengkth′), distance from one point on a wave to the same point on the next wave.

wedge (wej), a simple machine used to cut or split an object.

weight (wāt), the amount of force that gravity exerts on a mass.

weightlessness (wāt′lis nis), a condition in which the pull of gravity is balanced by other motion.

wheel and axle, simple machine that consists of a wheel and a shaft that turns with the wheel and supports its center.

work (wėrk), product of the force exerted on an object and the distance the force moves the object in the direction of the force; expressed in newton-meters or joules.

Index

D

Q

R

S

T

Acknowledgments

Pupil Editon interior design

Ligature, Inc.
Rosa + Wesley Design Associates

Unless otherwise acknowledged, all photographs are the property of ScottForesman. Unless otherwise acknowledged, all computer graphics by Ligature, Inc. Page abbreviations are as follows: **(T) top, (C) center, (B) bottom, (L) left, (R) right, (INS) inset.**

Module A

Photographs

Front & Back Cover: Background: David Scharf/Peter Arnold, Inc. Children's Photos: John Moore

Page A2(T) Uniphoto **A2(B)** The Granger Collection, New York **A6(TL)** S.Meola/Visuals Unlimited **A6(TR)** Dr.Jeremy Burgess/SPL/Photo Researchers **A7(T)** Dr.Jeremy Burgess/SPL/ Photo Researchers **A7(B)** David Scharf/Peter Arnold, Inc. **A12(T)** R.Weldon/Gemological Institute of America **A12(T INS)** R.Weldon/Gemological Institute of America **A12(B)** Art Resource **A12(B INS)** E.R.Degginger/Bruce Coleman, Inc. **A13(T)** Fred Ward/Black Star **A13(T INS)** DPI/Uniphoto **A13(B)** Charles Moore/Black Star **A13(B INS)** DPI/Uniphoto **A16-17** David Scharf/Peter Arnold, Inc. **A16(B INS)** David Scharf/Peter Arnold, Inc. **A23** Michel Viard/Peter Arnold, Inc. **A24(T)** Photo Courtesy of Scintag, Inc. **A24-25** Alfred Pasieka/SPL/Photo Researchers **A26** Jerry Howard/Stock Boston **A27(B)** Myrleen Ferguson/Photo Edit **A28** Richard Megna/Fundamental Photographs **A32(R)** Lawrence Berkeley Institute **A38(L)** The Granger Collection, New York **A38(R)** The Granger Collection, New York **A40** Yada Claassen/Stock Boston **A41(R)** John D. Cunningham/Visuals Unlimited **A49** Prof.J.Bories/CNRI/SPL/Photo Researchers **A50(T)** SIU/ Visuals Unlimited **A51-52** David M.Phillips/Visuals Unlimited **A55(L)** Dr.Jeremy Burgess/SPL/Photo Researchers **A55(R)** The Granger Collection, New York **A56** Biophoto Associates/SS/ Photo Researchers **A58(L)** M.Eichelberg/Visuals Unlimited **A58(R)** Kevin Vandiver **A59(T)** Dr.Jeremy Burgess/SPL/Photo Researchers **A59(B)** EM Unit, British Museum of Natural History/SPL/Photo Researchers **A60** CNRI/SPL/Photo Researchers **A61(TL)** Dr.Tony Brain & David Parker/SPL/Photo Researchers **A61(CL)** Dr.Tony Brain & David Parker/SPL/Photo Researchers **A61(CR)** Dr.Tony Brain & David Parker/SPL/ Photo Researchers **A61(R)** Dr.Tony Brain & David Parker/SPL/ Photo Researchers **A64(T)** David M.Phillips/Visuals Unlimited **A64(B)** Manfred Kage/Peter Arnold, Inc. **A65** Moredun Animal Health LTD/SPL/Photo Researchers **A66(T INS)** Chemical Design/SPL/Photo Researchers **A67(T INS)** Ed Reschke/Peter Arnold, Inc. **A67(B INS)** From CORPUSCLES by Marcel Bessis.Springer-Verlag, Berlin, Heidelberg, New York. © 1974 **A70** Dan McCoy/Rainbow **A71(L)** Manfred Kage/Peter Arnold, Inc. **A71(C)** Leonard Lessin/Peter Arnold, Inc.

Illustrations

Page A33-35 George Kelvin **A64** Biomedia **A65** Biomedia **A77** George Kelvin

Module B

Photographs

Front & Back Cover: Children's Photos: John Moore

Page B3 Prof.J.Bories/CNRI/SPL/Photo Researchers **B5** Dennis O`Clair/Tony Stone Worldwide **B16-17** Doug Arnand/Tony Stone Worldwide **B17** Robert Harding Picture Library Ltd., London **B18(L)** Science Museum, London **B32** Greg Vaughn/Tom Stack & Associates **B34(L)** Steve Elmore/The Stock Market **B34-35** Simon Nathan/The Stock Market **B36** American Honda Motor, Co, Inc **B37** G. Heisler/ The Image Bank **B37(T)** Reprinted with permission from POPULAR SCIENCE Magazine, copyright, 1991, TIMES MIRROR Magazines, Reprinted by permission, Los Angeles Times Syndicate. **B37(B)** G. Heisler/The Image Bank **B43(T)** The National Motor Museum at Beaulieu, England **B43(B)** Doug Menuez/Reportage **B44(T)** Culver Pictures **B44(B)** Culver Pictures **B45(C)** The National Motor Museum at Beaulieu, England **B45(T)** Vandystadt/ALLSPORT USA **B45(B)** Steven A Heller

Illustrations

Page B2 John Burgoyne **B17** John Burgoyne **B18** John Burgoyne **B19** John Burgoyne **B20-21** John Burgoyne **B32-33** Slug Signorino **B35** John Burgoyne **B58** George Kelvin **B61** Rich Lo

Module C

Photographs

Front & Back Cover: Background: Gray Mortimore/Tony Stone Worldwide Children's Photos: John Moore

Page C2(TL) Scott Bauer/Allen Laidman Photography **C13** Mike Powell/ALLSPORT USA **C14(T)** L. Tobey Sanford **C16(TL)** Culver Pictures **C16-17(T)** UPI/Bettmann **C16(B)** The Granger Collection, New York **C17(BR)** NASA **C23** Mark Seliger **C25(T)** The Granger Collection, New York **C25(B)** American Institute of Physics,/Niehls Bohr Library Photo Collection **C30** Manny Millan/Sports Illustrated for Kids **C32** Scott Bauer/Allen Laidman Photography **C40** R. Calentine/ Visuals Unlimited **C41** Bob Daemmrich/Stock Boston **C46(T)** Richard Hutchings/InfoEdit **C46(B)** Richard Hutchings/InfoEdit **C48** Louis Psihoyos/Matrix **C53** Focus On Sports **C54(L)** William Meyers/Third Coast Stock **C60(T)** NASA **C60(B)** NASA **C61** NASA **C68(T)** U.S. Space Camp(R) Photo **C68(C)** U.S. Space Camp(R) Photo **C68(B)** NASA **C69(C)** U.S. Space Camp(R) Photo **C69(T)** NASA **C69(B)** U.S. Space Camp(R) Photo **C74** Doug Hoke **C75** Doug Hoke

Illustrations

Page C3 Vincent Perez **C30** Vincent Perez **C32** Vincent Perez **C38-39** Vincent Perez **C41** Vincent Perez **C62** Jacque Auger **C66-67** Randy Verougstraete **C70-71** Randy Verougstraete **C77** Rich Lo

Module D

Photographs

Front & Back Cover: Background: Tony Stone Worldwide Children's Photos: John Moore

Page D2(T) Philip Jon Bailey/The Picture Cube **D3(C)** The Bettmann Archive **D8** Philip Jon Bailey/The Picture Cube **D9** Arnie Feinberg/The Picture Cube **D14(L)** F. Stuart Westmoreland/Tom Stack & Associates **D14(R)** Steve Ogden/Tom Stack & Associates **D15(L)** Courtesy American Petroleum Institute, Photographic and Film Archives **D15(R)** Courtesy Ford Motor Company **D18(T)** Doug Sokell/Visuals Unlimited **D18(B)** Jim Cornfield/West Light **D30** ASB Photos **D30-31** The Bettmann Archive **D31(TL)** Courtesy William Keck **D31(TR)** Wes Thompson/The Stock Market **D32(T)** NASA **D32(B)** Greg Vaughn/Tom Stack & Associates **D37(T)** Milt & Joan Mann/Cameramann International, Ltd. **D37(B)** Milt & Joan Mann/Cameramann International, Ltd. **D38(L)** Tony Stone Worldwide **D38(C)** COMSTOCK INC. **D38(R)** COMSTOCK INC. **D39(L)** Mark Stephenson/Pacific Rim/West Light **D39(C)** Varin J. P. Jacana/Photo Researchers **D39(R)** COMSTOCK INC. **D43** NASA **D51** Martin Rogers/Uniphoto **D70(R)** C.A. Morgan/Peter Arnold, Inc. **D71(T)** David R. Frazier Photolibrary **D76** Brent Jones/Scott, Foresman

Illustrations

Page D7 Clint Hansen **D28** Steve Fuller **D37** Teri McDermott **D41** Wild Onion Studio **D44-45** Wild Onion Studio **D60** Teri McDermott **D66-67** Walter Stuart **D68-69** Walter Stuart **D70** Wild Onion Studio **D77** George Kelvin

Module E

Photographs

Front & Back Cover: Background: Grant Heilman Photography Children's Photos: John Moore

Page E3(C) David Ryan/DDB Stock Photo **E8-9** Kent Wood/ Peter Arnold, Inc. **E9(T)** Grant Heilman Photography **E25** Michael Holford **E26** NASA **E27** Johnny Johnson/Alaska Photo/All Stock **E37** Alan Becker/The Image Bank **E39** Robert Frerck/Odyssey Productions, Chicago **E49(T)** John D. Cunningham/Visuals Unlimited **E55(T)** David Ryan/DDB Stock Photo **E55(B)** David Ryan/DDB Stock Photo **E57** Chris Alan Wilton/The Image Bank **E60(T)** Robert Fried/DDB Stock Photo **E60(BL)** Diane Johnson/Tony Stone Worldwide **E60-61(B)** Lawrence Ruggeri/Third Coast Stock **E61(T)** Bill Gallery/Stock Boston **E61(BR)** Erika Craddock/Tony Stone Worldwide **E68(T)** David Sarnoff Library **E68(B)** Science Museum, London **E69(TL)** Culver Pictures **E69(TR)** Science Museum, London **E69(BL)** Dan McCoy/Rainbow **E69(BR)** NASA **E74** Courtesy of Judy Leigh, Vanen Brook High School, Pepperell, MA. **E76** Mike Kelly/Scott, Foresman

Illustrations

Page E3 Dale Glasgow **E6** Mas Miyamoto **E9** Joe Rogers **E16** Carl Kock **E22** JAK Graphics **E33** Teri McDermott **E36** Teri McDermott **E38** Mas Miyamoto **E43** JAK Graphics **E43** Teri McDermott **E48** Susan Spellman **E56** Dale Glasgow **E70** Sharmen Liao **E77** George Kelvin

Module F

Photographs

Front & Back Cover: Background: W.H.Hodge/Peter Arnold, Inc. Children's Photos: John Moore

Page F3(T) Peter Menzel/Stock Boston **F2(B)** Robert Wagoner/ The Stock Market **F3L(C)** David Hanson/Tony Stone Worldwide **F7** Edward Farmer and Clarence Ryan/Washington State University, Institute of Biological Chemistry, Pullman **F8** Patrick Eggelston/Visuals Unlimited **F9** W.H. Hodge/Peter Arnold, Inc. **F12(L)** Grant Heilman Photography **F12(R)** Runk Schoenberger/Grant Heilman Photography **F13(T)** COMSTOCK INC. **F13(B)** C.P. Vance/Visuals Unlimited **F14** Hans Pfletschinger/Peter Arnold, Inc. **F17** Ray Atkeson/The Stock Market **F18(L)** Courtesy of the Trustees of the British Museum F18 UPI/Bettmann **F19(L)** UPI/Bettmann **F19(R)** Dick Thomas/Visuals Unlimited **F20** FPG **F21(T)** Steve Smith/West Light **F22(T)** Hans Pfletschinger/Peter Arnold, Inc. **F22(B)** D. Hagemann/Okapia/Photo Researchers **F23(T)** Science VU/Visuals Unlimited **F23(B)** Stephen Dalton/Photo Researchers **F31** Robert Wagoner/The Stock Market **F32-33** Bill Ross/West Light **F40** John Colwell/Grant Heilman Photography **F40-41** Swarthout/The Stock Market **F41(T)** Peter Menzel/Stock Boston **F43(T)** Grant Heilman Photography **F46(L)** by copyright permission of the AAAS. **F46(C)** Runk & Schoenberger/Grant Heilman Photography **F46(R)** by copyright permission of the AAAS. **F49** David M Dennis/Tom Stack & Associates **F50** David Hanson/Tony Stone Worldwide **F64** Roger Knutson/Soil Teq Inc **F65(T)** Linhoff/Soil Teq Inc **F65** Soil Teq Inc **F76** Audrey Trotman

Illustrations

Page F5 Carl Kock **F6** Carl Kock **F8** Ebet Dudley **F15** Carl Kock **F16** Laurie O'Keefe **F24-25** Nancy Lee Walter **F34** Wild Onion Studio **F35** Wild Onion Studio **F38** Wild Onion Studio **F40** Wild Onion Studio **F42** Wild Onion Studio **F56-57** Mas Miyamoto **F58-59** Mas Miyamoto **F59(T)** Carla Simmons **F60** Mas Miyamoto **F66-67** Simon Galkin **F68** George Kelvin **F70** Laurie O'Keefe